CRM Short Courses

The volumes in the **CRM Short Courses** series have a primarily instructional aim, focusing on presenting topics of current interest to readers ranging from graduate students to experienced researchers in the mathematical sciences. Each text is aimed at bringing the reader to the forefront of research in a particular area or field, and can consist of one or several courses with a unified theme. The inclusion of exercises, while welcome, is not strictly required. Publications are largely but not exclusively, based on schools, instructional workshops and lecture series hosted by, or affiliated with, the *Centre de Researches Mathématiques* (CRM). Special emphasis is given to the quality of exposition and pedagogical value of each text.

More information about this series at http://www.springer.com/series/15360

Yohann Le Floch

A Brief Introduction
to Berezin–Toeplitz Operators
on Compact Kähler Manifolds

CENTRE
DE RECHERCHES
MATHÉMATIQUES

Yohann Le Floch
IRMA
Université de Strasbourg
Strasbourg, France

ISSN 2522-5200 ISSN 2522-5219 (electronic)
CRM Short Courses
ISBN 978-3-030-06898-1 ISBN 978-3-319-94682-5 (eBook)
https://doi.org/10.1007/978-3-319-94682-5

Mathematics Subject Classification (2010): 53D50, 81S10, 81Q20, 32L05, 32Q15

This Springer imprint is published by the registered company Springer Nature Switzerland AG
The registered company address is: Gewerbestrasse 11, 6330 Cham, Switzerland

Preface

Berezin–Toeplitz operators naturally appear in quantum mechanics with compact classical phase space, when studying the semiclassical limit of the geometric quantisation procedure due to Kostant and Souriau. Lately, they have played a significant part in various areas of mathematics such as complex geometry, topological quantum field theory, integrable systems, and the study of some links between symplectic topology and quantum mechanics.

The aim of this book is to provide graduate and postgraduate students, as well as researchers, with a comprehensive introduction to these operators and their semiclassical properties, in the case of a Kähler phase space, with the least possible prerequisites. For this purpose, it contains a review of the relevant material from complex geometry, line bundles and integral operators on spaces of sections. The rest of the book is devoted to a proof of the main properties of Berezin–Toeplitz operators, which relies on the description of the asymptotic behaviour of the Bergman kernel. This description is taken for granted, but the reader can follow through the example of the projective space, for which this kernel is explicitly computed.

These notes were originally prepared for a course on Berezin–Toeplitz operators given with Leonid Polterovich at Tel Aviv University during the academic year 2015–2016. This course was mainly aimed at graduate students, and its goal was to prove the main properties of these operators and to present a few applications, following the ideas contained in the paper [20] by Laurent Charles and Leonid Polterovich. At the time, I decided to type some notes in order to provide the students with some examples, details and basics that we would not have time to cover during classes, but also to help myself in the organisation of my share of the lectures.

After the course was over, Leonid encouraged me to keep working on these notes and eventually try to publish them. After having been reluctant for some time, I finally let him convince me that it would be a good idea. Hence I came back to this project, added some material, and the present manuscript is an attempt at a clean version of these notes.

Acknowledgements I am extremely grateful to Leonid Polterovich for suggesting this course and for encouraging me to try to publish these notes and to Laurent Charles who explained some delicate details of their joint paper. I would like to thank all the people who attended the lectures at Tel Aviv University for their enthusiasm and numerous questions and remarks.

After a critical number of pages have been reached, writing an error-free text can be particularly difficult, if not impossible. This is why I would like to thank Ziv Greenhut for pointing out some mistakes in an earlier version of these notes, and why I am especially grateful to Vukašin Stojisavljević for devoting a lot of his time to reading very carefully the first half of this manuscript; his help has been extremely precious.

As this book is now becoming a reality, I would like to thank Véronique Hussin and André Montpetit of the CRM and Elizabeth Loew at Springer for their efficiency and professionalism. Having them as interlocutors has been a pleasure and has made my life easier with respect to this project. This is also a good occasion to thank the anonymous referees for their useful comments.

Finally, I am grateful to everyone who was involved in the final stages of production of the book at Scientific Publishing Services, and in particular to Shobana Ramamurthy for coordinating everything perfectly.

Strasbourg, France   Yohann Le Floch

Contents

Chapter 1
Introduction

Berezin–Toeplitz operators appear in the study of the semiclassical limit of the quantisation of compact symplectic manifolds. They were introduced by Berezin [5], their microlocal analysis was initiated by Boutet de Monvel and Guillemin [33], and they have been studied by many authors since, see for instance [8, 9, 14, 23, 30, 49]. This list is of course far from exhaustive, and the very nice survey paper by Schlichenmaier [43] gives a review of the Kähler case and contains a lot of additional useful references.

Besides consolidation of the theory, the past twenty years have seen the development of applications of these operators to various domains of mathematics and physics, such as topics in Kähler and algebraic geometry [22, 29, 41], topological quantum field theory [1, 18, 19, 32] or the study of integrable systems [7]. Moreover, they constitute a natural setting to investigate the connection between symplectic rigidity results on compact manifolds and their quantum consequences, and have recently been used to this effect by Charles and Polterovich [20, 21, 37, 38]. For all these reasons, their importance is now comparable to the one of pseudodifferential operators. Yet, while many textbooks on the latter are available, there is still, to our knowledge, no single place for a graduate student getting started on the subject of Berezin–Toeplitz operators to quickly learn the basic material that they need.

These notes are a modest attempt at filling this gap and are designed as an introduction to the case of compact Kähler manifolds, for which the constructions are simpler than in the general case. Before detailing their contents, let us explain how they have been built.

1.1 Overview of the Book

The philosophy of this book is to give a short and—hopefully—simple introduction to Berezin–Toeplitz operators on compact Kähler manifolds. Here, the word "simple" means that it has been written with the purpose of being understandable to, at least, graduate students; therefore, we have tried not to assume any know-

© Springer International Publishing AG, part of Springer Nature 2018
Y. Le Floch, *A Brief Introduction to Berezin–Toeplitz Operators on Compact Kähler Manifolds*, CRM Short Courses,
https://doi.org/10.1007/978-3-319-94682-5_1

ledge in the advanced material used throughout the different chapters. Thus, the minimal requirement is some acquaintance with the basics of differential geometry. Nevertheless, this does not mean that these notes are self-contained; despite all our efforts, we sometimes had to sacrifice completeness on the altar of concision. Furthermore, there is one major blackbox at the heart of these notes, namely the description of the asymptotics of the Bergman kernel (Theorem 7.2.1). The reason is that this result is quite involved, and presenting a proof would require space, the introduction of more advanced material and would go against the spirit of the present manuscript, which is to remain as short and non-technical as possible. Nonetheless, we will briefly sketch one of the most direct proofs we are aware of. Besides, one can directly start with the explicit form of the Bergman kernel in the case of $\mathbb{CP}^n$ (see Exercise 7.2.7), check that it satisfies all the conclusions of Theorem 7.2.1, and follow the rest of these notes with this particular example in mind.

The choice to focus on the case of Kähler manifolds is motivated by the fact that this is the setting in which the constructions are the simplest to explain, and that most of the usual examples belong to this class anyway. In particular, we will describe several examples on symplectic surfaces, which are automatically Kähler. We should nevertheless mention, for the interested reader, that there are several approaches to Berezin–Toeplitz operators in the general compact symplectic case: via almost-holomorphic sections and Fourier integral operators of Hermite type [10, 33, 44], via spin^c-Dirac operators [30] or, more recently, via a direct construction of a candidate for the Szegő projector [17]. We will not go any further in the discussion of the details of these constructions.

As regards the Kähler case, the quantisation procedure, named geometric quantisation, and due to Kostant [28] and Souriau [45], requires the existence of a certain complex line bundle over the manifold, called a prequantum line bundle. The Hilbert space of quantum states is then constructed as the space of holomorphic sections of some tensor power of this line bundle; in fact, the power in question serves as a semiclassical parameter, and we eventually consider a family of Hilbert spaces indexed by this power. Roughly speaking, the first half of this book is devoted to the construction and study of this family of Hilbert spaces; for a different point of view on this part, we recommend the excellent textbook by Woodhouse [47]. The second half deals with Berezin–Toeplitz operators, which are particular families of operators acting on these quantum spaces.

1.2 Contents

Since the aim of these notes is to give a brief introduction to the topic at hand, there is obviously a lot of material that has been left untouched. Our choice of the subjects to discuss or discard has been guided by two imperatives. Firstly, the notes follow the general guidelines of the course they were designed to accompany; namely, to introduce Berezin–Toeplitz operators on compact Kähler manifolds and state those of their properties which are needed to explain the main results from the three papers [20, 37, 38], with an audience knowing little—or even nothing—on the topic. This means that we have tried to make the exposition as clear as possible, and to refrain from going into full generality when this was not necessary.

Secondly, we have made the choice of focusing on the practical side of the subject, by devoting an important part of these notes to examples and useful tools. By doing so, we want to encourage the reader to immediately start playing with concrete Berezin–Toeplitz operators and check by themself that the properties stated in this book are satisfied by these examples.

One key feature that arose by taking into account these two aspects is the presence of exercises throughout the text. Again, we encourage the reader to try to solve these exercises, which constitute most of the time a simple verification that some notion or example has been understood, or a straightforward generalisation of some result. For these reasons, we do not provide with any solution to these exercises.

Keeping these guidelines in mind, let us now go to the heart of the subject, and explain further the general idea of the book. We want to quantise a phase space which is a compact Kähler manifold (M, ω, j), that is a compact manifold endowed with a symplectic form ω and a complex structure j, these two structures being compatible in some sense that we will not precise here. Roughly speaking, this means that we want to construct a Hilbert space $\mathcal{H}$, the space of quantum states, and to associate to each classical observable $f \in \mathcal{C}^\infty(M, \mathbb{R})$ a quantum observable, that is a self-adjoint operator $T(f) \in \mathcal{L}(\mathcal{H})$, in a way that respects a certain number of principles (note that we avoid discussing problems coming from the possible unboundedness of $T(f)$, which is fine since we will see that the relevant $\mathcal{H}$ will be finite-dimensional). More precisely, the map sending f to $T(f)$ must be linear, send the constant function equal to one to the identity of $\mathcal{H}$ and satisfy the famous correspondence principle, which states that the commutator $[T(f), T(g)]$ should be related to the quantum observable $T(\{f, g\})$ associated with the Poisson bracket of f and g. Before giving more precisions, let us insist on the fact that we want a semiclassical theory, so we want this construction to depend on Planck's constant $\hbar$ and to investigate the limit $\hbar \to 0$. Hence, what we really want is a family of Hilbert spaces $(\mathcal{H}_\hbar)_{\hbar>0}$ and a family of maps $f \mapsto T_\hbar(f)$, $\hbar > 0$. The geometric quantisation procedure requires the existence of an additional structure at the classical level, a holomorphic line bundle $L \to M$ with certain properties; the desired Hilbert spaces are then built as spaces of holomorphic sections of tensor powers of this line bundle. Hence, in this theory, what will play the role of $\hbar$ is the inverse of a positive integer k, and we will consider the family $(\mathcal{H}_k)_{k \geq 1}$ of spaces of holomorphic sections of $L^{\otimes k} \to M$; the semiclassical limit corresponds to $k \to +\infty$. The next step is to construct the family of maps $f \mapsto T_k(f)$, and the main objective of these notes is to prove that these maps satisfy the following properties as k goes to infinity:

(1) $\|T_k(f)\| \sim \|f\|_\infty$ (norm estimate),
(2) $T_k(f)T_k(g) \sim T_k(fg)$ (product estimate),
(3) $[T_k(f), T_k(g)] \sim (1/(ik))T_k(\{f, g\})$.

A rigorous version of these estimates was derived in the fundamental article [8], and the second and third were discussed in [23] together with the existence of a star product $\star$ such that $T_k(f)T_k(g) = T_k(f \star g)$ up to a remainder whose norm is small with respect to every negative power of k. One of the results in [31] is the explicit computation of the second-order term in the asymptotic expansion of $f \star g$. For all these results, f and g were assumed to be smooth; however, it is sometimes relevant to quantise non-smooth functions, and this case has been studied in [4].

Here, our goal is to prove more precise versions of the three facts above, where we add remainders on the right-hand sides, and we aim at describing these remainders in terms of f, g and their derivatives, following the article [20]. This goal will be reached as follows. Chapters 2 and 3 contain the minimal knowledge required to understand geometric quantisation, that is, respectively, some properties of Kähler manifolds and some facts about complex line bundles with connections; both chapters constitute quick overviews of the essential material, but are of course far from a complete treaty on these two topics. The reader who is already familiar with these two aspects may want to start with Chapter 4, where we describe the geometric quantisation procedure and investigate the first properties of the associated quantum spaces, such as the computation of their dimensions. In Chapter 5, we define Berezin–Toeplitz operators and state their properties, such as the estimate of their norm, and the behaviour of their compositions and commutators. The rest of the book is devoted to the proof of these three properties, based on the standard ansatz for the Schwartz kernel of the projector from the space of square integrable sections of the k-th tensor power of the prequantum line bundle to the space of its holomorphic sections. Consequently, Chapter 6 is devoted to a brief discussion of integral operators on spaces of sections and their kernels, before we describe the aforementioned Schwartz kernel in Chapter 7.

The reader must be warned once again that this description, stated in Theorem 7.2.1, is a highly non-trivial result, which led us to the choice of not proving it in these pages; this constitutes a major blackbox in these notes. The reason behind this choice is that we believe that adding a lengthy and technical proof would have reduced the clarity of the exposition, for almost no added value. Should the reader be interested in such a proof, we point to, and give a rough outline of, a very nice recent one [6], in Section 7.3; additionally, we very briefly explain in the Appendix how to derive Theorem 7.2.1 from a theorem of Boutet de Monvel and Sjöstrand on the Szegő projector of a strictly pseudoconvex domain (unfortunately, the latter is itself a difficult result). Alternatively, this kernel is explicitly computed in several examples, the most interesting ones being complex projective spaces, see Exercise 7.2.7. The computation in this case is accessible and can be easily checked by the reader, who can on the one hand obtain a complete derivation of all the results in these notes for projective spaces, and on the other hand convince themselves of the validity of the general result.

We investigate composition and commutators of Berezin–Toeplitz operators in Chapter 8. Finally, in Chapter 9, we explain how to estimate the norm of a Berezin–Toeplitz operator; to this effect, we introduce the so-called coherent states, and we use the rest of the chapter to discuss some nice properties of these states.

As a conclusion, we should warn the reader that they will not find anything new in these notes, but should rather see them as a convenient gathering of the folklore knowledge on the subject. We do not claim originality in any of the results contained in this manuscript.

1.3 Uncontents

As unbirthdays sometimes provide with more gifts and excitement than birthdays, the "uncontents" of these notes probably constitute the most interesting part of the topic, and it is worth mentioning the aspects that will not be evocated along these lines, if only to convince the interested reader that there is much more to learn about Berezin–Toeplitz operators. This will also allow us to point to a few references regarding these missing parts.

Perhaps the most important choice that we have made is to not talk about metaplectic correction. This first-order correction to quantisation is widely used, and in the context of geometric quantisation, it consists in working with holomorphic sections of $L^{\otimes k} \otimes \delta \to M$ instead of holomorphic sections of $L^{\otimes k} \to M$. Here $\delta \to M$ is a half-form bundle, that is a square root of the canonical bundle of M. Although this construction leads to nicer formulas, one example being the cancellation of the term of order k^{n-1} in the computation of the dimension of $\mathcal{H}_k$, the decision not to include it was not so complicated to make, because we felt that it would have led to a general obfuscation of the text and hindered the pedagogical writing that we have tried to use. Not only because replacing $L^{\otimes k}$ by $L^{\otimes k} \otimes \delta$ everywhere could have brought confusion to the reader, but also because such a half-form bundle may not exist globally over M, and this problem would have forced us to introduce some technical discussion. For more details on Berezin–Toeplitz operators within the framework of metaplectic correction, one can for instance look at the article by Charles on the subject [16].

A certain number of experts in Berezin–Toeplitz operators are used to working with circle bundles instead of line bundles. While we respect this choice, for semi-classical purposes, we have some good reasons to prefer using line bundles rather than circle bundles. However, a small number of proofs in these notes could have been simplified by adopting the circle bundle point of view, essentially in the section about the unitary evolution of Kostant–Souriau operators. We chose not to do so, since we realised that the gain would be small in comparison to the loss of efficiency induced by forcing the reader to digest a chapter on circle bundles. Nonetheless, for those who are interested in this aspect, and since we believe that it is useful to be able to easily pass from one theory to another, we have included an Appendix in which we compare the two points of view.

Besides these two major characters, there is a certain number of interesting topics that this book will not even allude to. In the product formula for Berezin–Toeplitz, one can go further than simply saying that the product $T_k(f)T_k(g)$ coincides with $T_k(fg)$ up to some small remainder. In fact, one can get a better approximation by comparing this product with $T_k\big(u(\cdot,k)\big)$ where u has a complete asymptotic expansion in negative powers of k (see, e.g. [23]). One can then talk about *a* subprincipal symbol, not *the* subprincipal symbol, since there are several choices of symbols. For more details about this and symbolic calculus, see for instance [14, 31]. A related question is the study of deformation quantisations on Kähler manifolds; it is discussed in [13, 23, 27, 42] for instance. We do not discuss Fourier integral operators in Kähler quantisation, and refer the reader to [15, 33, 48] for example. We do not mention the group theoretical aspects of geometric quantisation and Berezin–Toeplitz operators either, namely the quantisation of coadjoint orbits of compact Lie groups. Several references are available, but the original article by Kostant [28] constitutes a good starting place. Finally, as already explained, this book does not contain anything about the general symplectic case, and we invite the reader to have a look at the references listed above for this matter.

This list is of course not exhaustive (we could have cited spectral theory, trace formulae, etc.), and we hope that this introduction to Berezin–Toeplitz operators will give the reader the impulse to go through the looking glass and discover by themselves the wonders that lie on the other side.

Chapter 2
A Short Introduction to Kähler Manifolds

In this chapter, we recall some general facts about complex and Kähler manifolds. It is not an exhaustive list of such facts, but rather an introduction of objects and properties that we will need in the rest of the notes. The interested reader might want to take a look at some standard textbooks, such as [24, 35] for instance.

Let M be a smooth manifold (M will always be paracompact). The tangent (respectively cotangent) space at a point $m \in M$ will be denoted by $T_m M$ (respectively $T_m^* M$); the tangent (respectively cotangent) bundle will be denoted by TM (respectively $T^* M$). A vector field is a smooth section of the tangent bundle; the notation $\mathcal{C}^\infty(M, TM)$ will stand for the set of vector fields. Similarly, a differential form of degree p is a section of the exterior bundle $\Lambda^p(T^* M)$; we will use the notation $\Omega^p(M)$ for the set of degree p differential forms. We will write $i_X \alpha$ for the interior product of a vector field X with a differential form α.

2.1 Almost Complex Structures

Definition 2.1.1. An *almost complex structure* on M is a smooth field j of endomorphisms of the tangent bundle of M whose square is minus the identity:

$$\forall m \in M, \quad j_m^2 = -\operatorname{Id}_{T_m M}.$$

If such a structure exists, we say that (M, j) is an *almost complex manifold*.

By taking the determinant, we notice that if M is endowed with an almost complex structure, then its dimension is necessarily even. In what follows, we will denote this dimension by $2n$ with $n \geq 1$.

Example 2.1.2. Consider $M = \mathbb{R}^2$ with its standard basis, and let j be the endomorphism of $\mathbb{R}^2$ whose matrix in this basis is

$$J = \begin{pmatrix} 0 & -1 \\ 1 & 0 \end{pmatrix}.$$

© Springer International Publishing AG, part of Springer Nature 2018
Y. Le Floch, *A Brief Introduction to Berezin–Toeplitz Operators on Compact Kähler Manifolds*, CRM Short Courses,
https://doi.org/10.1007/978-3-319-94682-5_2

Then j is an almost complex structure on M; it corresponds to multiplication by i on $\mathbb{R}^2 \simeq \mathbb{C}$, $(x, y) \to x + iy$. More generally, the endomorphism of $\mathbb{R}^{2n}$ whose matrix in the standard basis is block diagonal with blocks as above is an almost complex structure on $\mathbb{R}^{2n}$.

This example is a particular case of a more general fact: if M is a complex manifold, i.e. a manifold modelled on $\mathbb{C}^n$ with holomorphic transition functions, then it has an almost complex structure. Indeed, let U be a trivialisation open set, and let $z_1, \ldots, z_n$ be holomorphic coordinates on U. For $\ell \in [\![1, n]\!]$, we define the functions $x_\ell = \Re(z_\ell)$ and $y_\ell = \Im(z_\ell)$. Then $(x_1, y_1, \ldots, x_n, y_n)$ are real coordinates on M, and

$$\forall \ell \in [\![1, n]\!], \qquad j\partial_{x_\ell} = \partial_{y_\ell}, \quad j\partial_{y_\ell} = -\partial_{x_\ell}$$

defines an almost complex structure j on M; it does not depend on the choice of local coordinates because the differentials of the transition functions are $\mathbb{C}$-linear isomorphisms, which means that they commute with this local j.

The converse is not true in general: an almost complex structure does not necessarily come from a structure of complex manifold. When it occurs, the almost complex structure is said to be *integrable*. We will state some integrability criterion later.

2.2 The Complexified Tangent Bundle

Given an almost complex manifold (M, j), we would like to diagonalise j; since it obviously has no real eigenvalue, we introduce the complexified tangent bundle $TM \otimes \mathbb{C}$ of M. We extend all endomorphisms of TM to its complexification by $\mathbb{C}$-linearity. Then we can decompose the complexified tangent bundle as the direct sum of the eigenspaces of j.

Lemma 2.2.1. *The complexified tangent bundle can be written as the direct sum*

$$TM \otimes \mathbb{C} = T^{1,0}M \oplus T^{0,1}M$$

where

$$T^{1,0}M := \ker(j - i\,\mathrm{Id}) = \{X - ijX \mid X \in TM\}$$

and

$$T^{0,1}M := \ker(j + i\,\mathrm{Id}) = \{X + ijX \mid X \in TM\} = \overline{T^{1,0}M}.$$

We will denote by $Y^{1,0}$ (respectively $Y^{0,1}$) the component in $T^{1,0}M$ (respectively $T^{0,1}M$) of an element Y of the complexified tangent bundle in this decomposition. We have that

$$Y^{1,0} = \frac{Y - ijY}{2}, \quad Y^{0,1} = \frac{Y + ijY}{2}$$

for such a Y.

Proof. Since $j^2 = -\,\mathrm{Id}$, j is diagonalisable over $\mathbb{C}$, with eigenvalues $\pm\mathrm{i}$:

$$TM \otimes \mathbb{C} = \ker(j - \mathrm{i}\,\mathrm{Id}) \oplus \ker(j + \mathrm{i}\,\mathrm{Id}).$$

Since these two eigenspaces correspond to complex conjugate eigenvalues, they are complex conjugate. Thus, it only remains to show that

$$\ker(j - \mathrm{i}\,\mathrm{Id}) = \{X - \mathrm{i}jX \mid X \in TM\}.$$

A simple computation shows that if $Y = X - \mathrm{i}jX$ with $X \in TM$, then $jY = \mathrm{i}Y$. Conversely, let $Z \in \ker(j - \mathrm{i}\,\mathrm{Id})$, and let us write $Z = X + \mathrm{i}Y$ with $X, Y \in TM$. From the equality

$$jX + \mathrm{i}jY = \mathrm{i}X - Y,$$

it follows, by identification of the real parts, that $Y = -jX$. $\qquad\square$

Let us assume that M is a complex manifold and that j is the associated complex structure introduced in the previous section. We consider some local complex coordinates $(z_1 = x_1 + \mathrm{i}y_1, \ldots, z_n = x_n + \mathrm{i}y_n)$, and define for $\ell \in [\![1, n]\!]$

$$\partial_{z_\ell} = \tfrac{1}{2}(\partial_{x_\ell} - \mathrm{i}\partial_{y_\ell}), \quad \partial_{\bar{z}_\ell} = \tfrac{1}{2}(\partial_{x_\ell} + \mathrm{i}\partial_{y_\ell});$$

then $(\partial_{z_\ell})_{1 \leq \ell \leq n}$ and $(\partial_{\bar{z}_\ell})_{1 \leq \ell \leq n}$ are local bases of $T^{1,0}M$ and $T^{0,1}M$, respectively.

The following statement gives a necessary and sufficient condition for an almost complex structure to induce a genuine complex structure. Let us recall that a distribution $E \subset TM$ is integrable if and only if for any two vector fields $X, Y \in E$, the Lie bracket $[X, Y]$ belongs to E (this is actually equivalent to the usual definition as a consequence of the Frobenius integrability theorem, but we take it as a definition to simplify).

Theorem 2.2.2 (The Newlander–Nirenberg theorem). *Let (M, j) be an almost complex manifold. Then j comes from a complex structure if and only if the distribution $T^{1,0}M$ is integrable.*

A proof of this standard but rather involved result can be found in [25, Section 5.7] for instance.

2.3 Decomposition of Forms

By duality, the decomposition $TM \otimes \mathbb{C} = T^{1,0}M \oplus T^{0,1}M$ induces a decomposition of the complexified cotangent bundle:

$$T^*M \otimes \mathbb{C} = (T^*M)^{1,0} \oplus (T^*M)^{0,1}$$

where

$$(T^*M)^{1,0} = \{\alpha \in T^*M \mid \forall X \in T^{0,1}M, \alpha(X) = 0\},$$

and $(T^*M)^{0,1}$ is defined in the same way, replacing $T^{0,1}M$ by $T^{1,0}M$. Similarly to Lemma 2.2.1, we have the following description.

Lemma 2.3.1. *We have that*

$$(T^*M)^{1,0} = \{\alpha - \mathrm{i}\alpha \circ j \mid \alpha \in T^*M\}, \quad (T^*M)^{0,1} = \overline{(T^*M)^{1,0}}.$$

It is well-known that the exterior algebra of a direct sum of two vector spaces is isomorphic to the tensor product of both exterior algebras of the vector spaces, and that this isomorphism respects the grading. Consequently, we have that

$$\Lambda^k(T^*M) \otimes \mathbb{C} \simeq \bigoplus_{\ell=0}^{k} (\Lambda^{\ell,0}M \otimes \Lambda^{0,k-\ell}M)$$

with $\Lambda^{p,0}M := \Lambda^p\big((T^*M)^{1,0}\big)$ and $\Lambda^{0,q}M := \Lambda^q\big((T^*M)^{0,1}\big)$. This can be written as

$$\Lambda^k(T^*M) \otimes \mathbb{C} \simeq \bigoplus_{\substack{p,q \in \mathbb{N} \\ p+q=k}} \Lambda^{p,q}M$$

with $\Lambda^{p,q}M := \Lambda^{p,0}M \otimes \Lambda^{0,q}M$. Therefore, this induces a decomposition of the space of k-forms:

$$\Omega^k(M) \otimes \mathbb{C} = \bigoplus_{\substack{p,q \in \mathbb{N} \\ p+q=k}} \Omega^{p,q}(M)$$

where $\Omega^{p,q}(M)$ is the space of smooth sections of $\Lambda^{p,q}M$. An element of $\Omega^{p,q}(M)$ will be called a (p,q)-form. These forms can be characterised in the following way.

Lemma 2.3.2. *A k-form α belongs to $\Omega^{k,0}(M)$ if and only if for every vector field $X \in \mathcal{C}^\infty(M, T^{0,1}M)$, $i_X\alpha = 0$. More generally, a k-form α belongs to $\Omega^{p,q}(M)$ with $p + q = k$, $q \neq k$, if and only if for any $q + 1$ vector fields $X_1, \dots, X_{q+1} \in \mathcal{C}^\infty(M, T^{0,1}M)$, $i_{X_1} \dots i_{X_{q+1}}\alpha = 0$.*

By applying complex conjugation, we deduce from this result that a k-form α belongs to $\Omega^{p,q}(M)$ with $p + q = k$, $p \neq k$, if and only if for any $p + 1$ vector fields $Y_1, \dots, Y_{p+1} \in \mathcal{C}^\infty(M, T^{1,0}M)$, $i_{X_1} \dots i_{X_{p+1}}\alpha = 0$.

Proof. Let $\alpha \in \Omega^{k,0}(M)$. We can write α locally as a sum of terms of the form

$$\alpha_\ell = \beta_1 \wedge \dots \wedge \beta_k$$

with $\beta_1, \dots, \beta_k \in \Omega^{1,0}(M)$. If $X \in \mathcal{C}^\infty(M, T^{0,1}M)$, by using the formula

$$i_X(\gamma \wedge \delta) = (i_X\gamma) \wedge \delta + (-1)^{\deg \gamma}\gamma \wedge (i_X\delta)$$

for differential forms γ, δ, and the fact that $\beta_j(X) = 0$, we obtain that $i_X\alpha = 0$. Conversely, let $\alpha \in \Omega^k(M) \otimes \mathbb{C}$ whose interior product with every $X \in \mathcal{C}^\infty(M, T^{0,1}M)$ vanishes. We write as

$$\alpha = \alpha^{(k,0)} + \alpha^{(k-1,1)} + \cdots + \alpha^{(0,k)}$$

the decomposition of α in the direct sum $\Omega^k(M) = \Omega^{k,0}(M) \oplus \cdots \oplus \Omega^{0,k}(M)$. For $X \in \mathcal{C}^\infty(M, T^{0,1}M)$, one has

$$0 = i_X \alpha = i_X \alpha^{(k-1,1)} + \cdots + i_X \alpha^{(0,k)}$$

since $i_X \alpha^{(k,0)} = 0$ by the first part of the proof. It is easy to check that $i_X \alpha^{(k-p,p)}$ belongs to $\Omega^{k-p,p-1}(M)$ for $1 \le p \le k$. Therefore, the previous equality yields that $i_X \alpha^{(k-p,p)} = 0$ for every $p \in [\![1,k]\!]$. Now, we take a local basis $\beta_1, \ldots \beta_n$ of $(T^*M)^{1,0}$ and write

$$\alpha^{(k-p,p)} = \sum_{\substack{L=\{\ell_1,\ldots,\ell_{k-p}\}\subset[\![1,n]\!]\\ \ell_1<\cdots<\ell_{k-p}}} \sum_{\substack{M=\{m_1,\ldots,m_p\}\subset[\![1,n]\!]\\ m_1<\cdots<m_p}} f_{L,M}\beta_{\ell_1} \wedge \cdots \wedge \beta_{\ell_{k-p}} \wedge \bar{\beta}_{m_1} \wedge \cdots \wedge \bar{\beta}_{m_p}$$

for some smooth functions $f_{L,M}$. Then

$$i_X \alpha^{(k-p,p)} = \sum_L \sum_M \sum_{r=1}^p \pm f_{L,M}\bar{\beta}_{\ell_r}(X)\beta_{\ell_1} \wedge \cdots \wedge \beta_{\ell_{k-p}} \wedge \bar{\beta}_{m_1} \wedge \cdots \wedge \bar{\beta}_{m_{r-1}} \wedge \bar{\beta}_{m_{r+1}} \wedge \cdots \wedge \bar{\beta}_{m_p},$$

thus $f_{L,M}\bar{\beta}_{m_r}(X) = 0$ for every L, M, m_r and every $X \in \mathcal{C}^\infty(M, T^{0,1}M)$, which finally yields $\alpha^{(k-p,p)} = 0$. Therefore $\alpha = \alpha^{(k,0)}$ belongs to $\Omega^{k,0}(M)$.

The second statement can be proved by induction on q (the first statement is the $q = 0$ case). $\qquad\square$

We would like to understand the action of the exterior derivative (extended by $\mathbb{C}$-linearity) with respect to this decomposition. It turns out that it behaves "nicely" if and only if j is induced by a structure of complex manifold on M. Before explaining this, let us introduce one more object.

Definition 2.3.3. The *Nijenhuis tensor* N_j of j is defined as follows: for any vector fields $X, Y \in \mathcal{C}^\infty(M, TM)$,

$$N_j(X,Y) = [X,Y] + j[jX,Y] + j[X,jY] - [jX,jY].$$

This tensor allows one to express the integrability condition in the Newlander–Nirenberg theorem in a more algebraic way.

Proposition 2.3.4. *Let (M,j) be an almost complex manifold. The following assertions are equivalent:*

(1) *j comes from a complex structure,*
(2) *$d(\Omega^{1,0}(M)) \subset \Omega^{2,0}(M) \oplus \Omega^{1,1}(M)$,*
(3) *$\forall p, q \in \mathbb{N},\ d(\Omega^{p,q}(M)) \subset \Omega^{p+1,q}(M) \oplus \Omega^{p,q+1}(M)$,*
(4) *$N_j = 0$.*

Proof. (1) $\Leftrightarrow$ (4): the Newlander–Nirenberg theorem states that j comes from a complex structure if and only if $[\mathcal{C}^\infty(M, T^{1,0}M), \mathcal{C}^\infty(M, T^{1,0}M)] \subset \mathcal{C}^\infty(M, T^{1,0}M)$. So let $X, Y \in \mathcal{C}^\infty(M, TM)$; a straightforward computation yields

$$[X - \mathrm{i}jX, Y - \mathrm{i}jY] = [X, Y] - [jX, jY] - \mathrm{i}[X, jY] - \mathrm{i}[jX, Y],$$

which implies that

$$[X - \mathrm{i}jX, Y - \mathrm{i}jY]^{0,1} = \tfrac{1}{2}\big(N_j(X, Y) + \mathrm{i}jN_j(X, Y)\big).$$

Therefore, $[X - \mathrm{i}jX, Y - \mathrm{i}jY]$ belongs to $\mathcal{C}^\infty(M, T^{1,0}M)$ if and only if $N_j(X, Y) = 0$.
  (1) $\Leftrightarrow$ (2): let $\alpha \in \Omega^{1,0}(M)$, and let $X, Y \in \mathcal{C}^\infty(M, T^{0,1}M)$. Then

$$\mathrm{d}\alpha(X, Y) = \mathcal{L}_X\big(\alpha(Y)\big) - \mathcal{L}_Y\big(\alpha(X)\big) - \alpha([X, Y]) = -\alpha([X, Y])$$

because $\alpha(X) = 0 = \alpha(Y)$ by definition of $\Omega^{1,0}(M)$ (here $\mathcal{L}_X$ stands for the Lie derivative with respect to X). Therefore, $\mathrm{d}\alpha(X, Y) = 0$ if and only if $[X, Y]^{1,0} \in \ker \alpha$. This means that $\mathrm{d}\big(\Omega^{1,0}(M)\big) \subset \Omega^{2,0}(M) \oplus \Omega^{1,1}(M)$ if and only if for any $X, Y \in \mathcal{C}^\infty(M, T^{0,1}M)$, $[X, Y]^{1,0} = 0$, i.e. $[X, Y]$ belongs to $\mathcal{C}^\infty(M, T^{0,1}M)$.
  (2) $\Leftrightarrow$ (3): the implication (3) $\Rightarrow$ (2) is obvious. Thus, let us assume that statement (2) holds. By complex conjugation, this implies that

$$\mathrm{d}\big(\Omega^{0,1}(M)\big) \subset \Omega^{1,1}(M) \oplus \Omega^{0,2}(M)$$

as well. Let $\gamma \in \Omega^{p,q}(M)$; we can write locally γ as a sum of elements $\tilde{\gamma}$ of the form

$$\tilde{\gamma} = \alpha_1 \wedge \cdots \wedge \alpha_p \wedge \beta_1 \wedge \cdots \wedge \beta_q$$

with $\alpha_1, \ldots, \alpha_p \in \Omega^{1,0}(M)$, $\beta_1, \ldots, \beta_q \in \Omega^{0,1}(M)$. Then by the Leibniz rule for forms

$$\mathrm{d}\tilde{\gamma} = \mathrm{d}\alpha_1 \wedge \hat{\gamma}_{\alpha_1} + \cdots + \mathrm{d}\alpha_p \wedge \hat{\gamma}_{\alpha_p} + \mathrm{d}\beta_1 \wedge \hat{\gamma}_{\beta_1} + \cdots + \mathrm{d}\beta_q \wedge \hat{\gamma}_{\beta_q}.$$

where $\hat{\gamma}_{\alpha_j} = \alpha_1 \wedge \cdots \wedge \alpha_{j-1} \wedge \alpha_{j+1} \wedge \cdots \wedge \alpha_p \wedge \beta_1 \wedge \cdots \wedge \beta_q$ and $\hat{\gamma}_{\beta_j}$ is defined in the same way. In particular, $\hat{\gamma}_{\alpha_j}$ belongs to $\Omega^{p-1,q}(M)$. Moreover, since α_j is a $(1,0)$-form, $\mathrm{d}\alpha_j$ belongs to $\Omega^{2,0}(M) \oplus \Omega^{1,1}(M)$; therefore $\mathrm{d}\alpha_j \wedge \hat{\gamma}_{\alpha_j}$ belongs to $\Omega^{p+1,q}(M) \oplus \Omega^{p,q+1}(M)$. It follows from a similar reasoning that $\mathrm{d}\beta_k \wedge \hat{\gamma}_{\beta_k}$ also belongs to this direct sum. Consequently, $\mathrm{d}\tilde{\gamma}$ belongs to $\Omega^{p+1,q}(M) \oplus \Omega^{p,q+1}(M)$, and so does $\mathrm{d}\gamma$. $\square$

  Observe that, as a consequence of this result, an almost complex structure on a surface always comes from a complex structure. Indeed, if M is a surface, then $\Omega^{2,0}(M) = \Omega^{0,2}(M) = \{0\}$, and therefore the exterior derivative of a $(1,0)$-form always lies in $\Omega^{1,1}(M)$.

2.4 Complex Manifolds

Let us now assume that M is a complex manifold and that j is the induced almost complex structure. Let $(z_k = x_k + iy_k)_{1 \leq k \leq n}$ be some local complex coordinates on an open subset $U \subset M$. We get complex-valued forms

$$\mathrm{d}z_k = \mathrm{d}x_k + i\mathrm{d}y_k \in \Omega^{1,0}(U), \quad \mathrm{d}\bar{z}_k = \mathrm{d}x_k - i\mathrm{d}y_k \in \Omega^{0,1}(U)$$

which form local bases $(\mathrm{d}z_1, \ldots, \mathrm{d}z_n)$, $(\mathrm{d}\bar{z}_1, \ldots, \mathrm{d}\bar{z}_n)$ of $\Omega^{1,0}(M)$ and $\Omega^{0,1}(M)$, respectively; $(\mathrm{d}z_1, \ldots, \mathrm{d}z_n, \mathrm{d}\bar{z}_1, \ldots, \mathrm{d}\bar{z}_n)$ is a local basis of $\Omega^1(M) \otimes \mathbb{C}$ which is dual to the local basis $(\partial_{z_1}, \ldots, \partial_{z_n}, \partial_{\bar{z}_1}, \ldots, \partial_{\bar{z}_n})$ introduced above. Therefore, a local basis of $\Omega^{p,q}(M)$ is given by the family

$$(\mathrm{d}z_{k_1} \wedge \cdots \wedge \mathrm{d}z_{k_p} \wedge \mathrm{d}\bar{z}_{\ell_1} \wedge \cdots \wedge \mathrm{d}\bar{z}_{\ell_q})_{1 \leq k_1 < \cdots < k_p, \ell_1 < \cdots < \ell_q \leq n}.$$

This immediately provides one with another proof of the fact that, in this case, the image of $\Omega^{p,q}(M)$ by the exterior derivative is included in $\Omega^{p+1,q}(M) \oplus \Omega^{p,q+1}(M)$. Because of this fact, we can write $\mathrm{d} = \partial + \bar{\partial}$ where

$$\partial \colon \Omega^{p,q}(M) \to \Omega^{p+1,q}(M), \quad \bar{\partial} \colon \Omega^{p,q}(M) \to \Omega^{p,q+1}(M).$$

The operators $\partial, \bar{\partial}$ satisfy the Leibniz rule

$$\partial(\alpha \wedge \beta) = \partial\alpha \wedge \beta + (-1)^{\deg(\alpha)} \alpha \wedge \partial\beta, \quad \bar{\partial}(\alpha \wedge \beta) = \bar{\partial}\alpha \wedge \beta + (-1)^{\deg(\alpha)} \alpha \wedge \bar{\partial}\beta,$$

which we can prove by writing the Leibniz rule for d and identifying the types.

Let $(U_k, \varphi_k)_{k \in I}$ be a holomorphic atlas of M. A function $f \colon M \to \mathbb{C}$ is called holomorphic if and only if for every $k \in I$, the function $f \circ \varphi_k^{-1} \colon \mathbb{C}^n \to \mathbb{C}$ is holomorphic.

Lemma 2.4.1. *Let $f \colon M \to \mathbb{C}$ be a smooth function. The following statements are equivalent*

(1) *f is holomorphic,*
(2) *for every $Z \in \mathcal{C}^\infty(M, T^{0,1}M)$, $\mathcal{L}_Z f = 0$,*
(3) *$\bar{\partial} f = 0$.*

Proof. The equivalence of the last two statements is clear because $\bar{\partial} f = 0$ is equivalent to the fact that $\mathrm{d}f$ belongs to $\Omega^{1,0}(M)$. Now, let $(z_1, \ldots, z_n)$ be the local complex coordinates defined by φ_k; then f is holomorphic if and only if

$$\forall \ell \in [\![1, n]\!], \quad \frac{\partial f}{\partial \bar{z}_\ell} = 0$$

in these coordinates. This is equivalent to saying that $\mathrm{d}f(\partial_{\bar{z}_\ell}) = 0$ for every ℓ; since $(\partial_{\bar{z}_\ell})_{1 \leq \ell \leq n}$ is a local basis of $T^{0,1}M$, this amounts to $\mathrm{d}f \in \Omega^{1,0}(M)$, which in turn is equivalent to $\bar{\partial} f = 0$. $\qquad\square$

Lemma 2.4.2. *The following identities hold:*

$$\partial^2 = 0, \quad \partial\bar{\partial} + \bar{\partial}\partial = 0, \quad \bar{\partial}^2 = 0.$$

Proof. This follows from the equality

$$0 = \mathrm{d}^2 = \partial^2 + \partial\bar{\partial} + \bar{\partial}\partial + \bar{\partial}^2$$

and the fact that $\partial^2 \colon \Omega^{p,q}(M) \to \Omega^{p+2,q}(M)$, $\partial\bar{\partial} + \bar{\partial}\partial \colon \Omega^{p,q}(M) \to \Omega^{p+1,q+1}(M)$ and $\bar{\partial}^2 \colon \Omega^{p,q}(M) \to \Omega^{p,q+2}(M)$.

Following the standard terminology for the exterior derivative, we say that a complex-valued form α is $\bar{\partial}$-*closed* if $\bar{\partial}\alpha = 0$, and $\bar{\partial}$-*exact* if there exists a differential form β such that $\alpha = \bar{\partial}\beta$. The operator $\bar{\partial}$ defines a cohomology, called Dolbeault cohomology. The cohomology groups are the quotients of $\bar{\partial}$-closed forms by $\bar{\partial}$-exact forms:

$$H^{p,q}(M) = \ker\bigl(\bar{\partial}_{|\Omega^{p,q}(M)}\bigr)\big/\bar{\partial}\bigl(\Omega^{p,q-1}(M)\bigr).$$

The following result is an analogue of the Poincaré lemma for the exterior derivative.

Lemma 2.4.3 (Dolbeault–Grothendieck lemma, or $\bar{\partial}$-Poincaré lemma). *A $\bar{\partial}$-closed form is locally $\bar{\partial}$-exact.*

For a proof, we refer the reader to standard textbooks, for instance [24, Proposition 1.3.8]. This result can be used to prove the following property of the operator $\mathrm{i}\partial\bar{\partial}$, which will be useful later.

Lemma 2.4.4 (The $\mathrm{i}\partial\bar{\partial}$-lemma). *Let $\alpha \in \Omega^{1,1}(M)$ be a differential form of type $(1,1)$. Then α is closed and real-valued (i.e., $\alpha \in \Omega^{1,1}(M) \cap \Omega^2(M)$) if and only if every point $m \in M$ has an open neighbourhood U such that $\alpha = \mathrm{i}\partial\bar{\partial}\phi$ over U for some $\phi \in \mathcal{C}^\infty(U,\mathbb{R})$.*

Proof. Assume that $\alpha = \mathrm{i}\partial\bar{\partial}\phi$ over some open subset $U \subset M$ for some $\phi \in \mathcal{C}^\infty(U,\mathbb{R})$. Then

$$\mathrm{d}\alpha = \mathrm{i}(\partial^2\bar{\partial}\phi + \bar{\partial}\partial\bar{\partial}\phi) = -\mathrm{i}\partial\bar{\partial}^2\phi = 0,$$

which proves that α is closed. Moreover,

$$\bar{\alpha} = -\mathrm{i}\bar{\partial}\partial\bar{\phi} = -\mathrm{i}\bar{\partial}\partial\phi = \mathrm{i}\partial\bar{\partial}\phi = \alpha,$$

thus α is real-valued.

Conversely, assume that α is closed and real-valued. From the usual Poincaré lemma, there exists locally a real-valued one-form β such that $\alpha = \mathrm{d}\beta$. From the equality

$$\alpha = \mathrm{d}\beta = \partial\beta^{(1,0)} + \bar{\partial}\beta^{(1,0)} + \partial\beta^{(0,1)} + \bar{\partial}\beta^{(0,1)},$$

we deduce that $\alpha = \bar{\partial}\beta^{(1,0)} + \partial\beta^{(0,1)}$ and $\bar{\partial}\beta^{(0,1)} = 0$. Thanks to the Dolbeault–Grothendieck lemma, we can find a local function f such that $\beta^{(0,1)} = \bar{\partial}f$. Since β is real-valued, the components of β satisfy

$$\beta^{(1,0)} = \overline{\beta^{(0,1)}} = \partial \bar{f}.$$

We finally obtain that

$$\alpha = \bar{\partial}\partial \bar{f} + \partial\bar{\partial}f = \partial\bar{\partial}(f - \bar{f}) = i\partial\bar{\partial}\phi$$

with $\phi = 2\Im(f)$. $\square$

2.5 Kähler Manifolds

Let (M, j) be an almost complex manifold.

Definition 2.5.1. A Riemannian metric g on M is said to be *compatible* with j if

$$g(jX, jY) = g(X, Y)$$

for every $X, Y \in TM$.

Every almost complex manifold can be equipped with a compatible Riemannian metric. Indeed, take any Riemannian metric g on M and define

$$h(X, Y) := g(X, Y) + g(jX, jY)$$

for every $X, Y \in TM$; then h is compatible with j. Given a compatible Riemannian metric g on (M, j), one defines its *fundamental form* as

$$\omega(X, Y) := g(jX, Y)$$

for every $X, Y \in TM$.

Lemma 2.5.2. *The fundamental form ω is a real $(1, 1)$-form, i.e., it belongs to $\Omega^{1,1}(M) \cap \Omega^2(M)$.*

Proof. Firstly, we check that ω belongs to $\Omega^2(M)$:

$$\omega(Y, X) = g(jY, X) = g(j^2Y, jX) = -g(Y, jX) = -g(jX, Y) = -\omega(X, Y)$$

for $X, Y \in TM$. Secondly, to prove that ω is of type $(1, 1)$, it is enough, by Lemma 2.3.2, to show that it vanishes when applied to a pair of elements of $T^{1,0}M$. Therefore, let $X, Y \in \mathcal{C}^\infty(M, TM)$; then

$$\omega(X - ijX, Y - ijY) = \omega(X, Y) - i\omega(X, jY) - i\omega(jX, Y) - \omega(jX, jY).$$

But on the one hand

$$\omega(jX, jY) = g(j^2X, jY) = -g(X, jY) = -g(jX, j^2Y) = g(jX, Y) = \omega(X, Y)$$

and on the other hand

$$\omega(jX,Y) = g(j^2X,Y) = -g(X,Y) = -g(jX,jY) = -\omega(X,jY).$$

Consequently, $\omega(X - ijX, Y - ijY) = 0$. $\square$

To illustrate this, let us assume for a moment that M is a complex manifold and that j is the induced almost complex structure. We choose some local holomorphic coordinates $(z_1, \ldots, z_n)$, and define the function

$$h_{\ell,m} := g(\partial_{z_\ell}, \partial_{\bar{z}_m})$$

where g has been extended to $TM \otimes \mathbb{C}$ by $\mathbb{C}$-bilinearity (and not sesquilinearity!). One can check that

$$\omega = i \sum_{\ell,m=1}^{n} h_{\ell,m} \, dz_\ell \wedge d\bar{z}_m$$

in these coordinates.

Let (M,j) be an almost complex manifold, and let g be a compatible Riemannian metric. Since j is an isomorphism and g is non-degenerate, it is clear that ω is non-degenerate. Hence, if it is closed, it is a symplectic form.

Definition 2.5.3. A compatible Riemannian metric on an almost complex manifold is called a *Kähler metric* if j is integrable and the fundamental form ω is closed. A *Kähler manifold* (M,j,g) is an almost complex manifold (M,j) endowed with a Kähler metric g.

In this case, the fundamental form is a symplectic form on M. By Lemma 2.4.4, near each point $p \in M$, there exists a real-valued smooth function ϕ such that

$$\omega = i\partial\bar{\partial}\phi.$$

This function ϕ is called a *Kähler potential*. In local coordinates, this gives

$$h_{\ell,m} = \frac{\partial^2 \phi}{\partial z_\ell \partial \bar{z}_m}$$

which means that the metric is determined locally by the Kähler potential.

In what follows, we will be more interested in the symplectic point of view. So let us start with a symplectic manifold (M,ω).

Definition 2.5.4. An almost complex structure j on M is said to be *compatible* with ω if

$$\omega(jX,jY) = \omega(X,Y)$$

for any $X,Y \in TM$ and

$$\omega(X,jX) > 0$$

for every $X \neq 0 \in TM$.

One readily checks that, given a compatible almost complex structure j on (M, ω), the spaces $T^{1,0}M$ and $T^{0,1}M$ are Lagrangian (when we extend ω to $TM \otimes \mathbb{C}$ by $\mathbb{C}$-bilinearity).

Assume that M is a complex manifold endowed with the induced complex structure j, and that ω belongs to $\Omega^{1,1}(M)$. In local coordinates $(z_\ell)_{1 \leq \ell \leq n}$, the symplectic form is of the form

$$\omega = \mathrm{i} \sum_{\ell, m = 1}^{n} h_{\ell, m} \, \mathrm{d}z_\ell \wedge \mathrm{d}\bar{z}_m$$

for some smooth functions $h_{\ell, m}$, $1 \leq \ell, m \leq n$. Then ω is compatible with j if and only if all the matrices $\left(h_{\ell, m}(p)\right)_{1 \leq \ell, m \leq n}$, $p \in M$ coming from such local expressions are positive definite Hermitian matrices.

A symplectic manifold always has a compatible almost complex structure. Indeed, take any Riemannian metric g on M. By the Riesz representation theorem, we have two isomorphisms

$$\widetilde{\omega} \colon TM \to T^*M, \quad X \mapsto i_X \omega \quad \text{and} \quad \tilde{g} \colon TM \to T^*M, \quad X \mapsto g(X, \cdot).$$

Consider $a = \tilde{g}^{-1} \circ \widetilde{\omega} \colon TM \to TM$; it is an isomorphism, which is moreover anti-symmetric, in the sense that $a^* = -a$ (a^* is the adjoint of a with respect to g). Indeed,

$$g(aX, Y) = \omega(X, Y) = -\omega(Y, X) = -g(aY, X) = g(X, -aY)$$

for any $X, Y \in TM$. Let

$$a = j(a^*a)^{1/2}$$

be the polar decomposition of a; j is unitary (with respect to g).

Lemma 2.5.5. *j is an almost complex structure which is compatible with ω.*

Proof. On the one hand, since j is unitary, $j^*j = \mathrm{Id}_{TM}$. On the other hand, since $(a^*a)^{1/2}$ is an isomorphism commuting with a and a is anti-symmetric, we have

$$j^* = \left(a(a^*a)^{-1/2}\right)^* = (a^*a)^{-1/2}a^* = -(a^*a)^{-1/2}a = -a(a^*a)^{-1/2} = -j,$$

thus $j^2 = -\mathrm{Id}_{TM}$. It remains to check the compatibility between j and ω. Firstly, we have that

$$\omega(jX, jY) = g(ajX, jY) = g(jaX, jY) = g(aX, j^*jY) = g(aX, Y) = \omega(X, Y)$$

for any $X, Y \in TM$. Secondly, for every $X \neq 0 \in TM$,

$$\omega(X, jX) = g(aX, jX) = g(j^*aX, X) = g\left((a^*a)^{1/2}X, X\right) > 0$$

because $(a^*a)^{1/2}$ is positive definite. $\square$

Observe that, given a symplectic form ω and a compatible almost complex structure j, the formula

$$g(X, Y) := \omega(X, jY)$$

defines a Riemannian metric on M, which is compatible with j and whose fundamental form is equal to ω. The latter is closed by definition; therefore we obtain an equivalent definition of Kähler manifolds.

Proposition 2.5.6. *A symplectic manifold (M, ω) is a Kähler manifold if and only if there exists an almost complex structure j which is compatible with ω and integrable.*

Note that an orientable surface is always a Kähler manifold. Indeed, by the discussion above, it can be endowed with an almost complex structure compatible with the symplectic (volume) form. But as we noticed earlier, an almost complex structure on a surface is always integrable.

Example 2.5.7. On $\mathbb{C}$ with its standard complex structure, the standard symplectic form

$$\omega = \frac{\mathrm{i}}{2} \sum_{\ell=1}^{n} \mathrm{d}z_\ell \wedge \mathrm{d}\bar{z}_\ell$$

is the fundamental form of the Kähler metric given by the standard scalar product on $\mathbb{R}^{2n}$. There is a globally defined Kähler potential given by

$$\phi(z_1, \ldots, z_n, \bar{z}_1, \ldots, \bar{z}_n) = \frac{1}{2} \sum_{\ell=1}^{n} |z_\ell|^2.$$

Example 2.5.8 (*The unit disc*). On the open unit disc $\mathbb{D}^n \subset \mathbb{C}^n$ (still with standard complex structure), we consider the function

$$\phi(z_1, \ldots, z_n, \bar{z}_1, \ldots, \bar{z}_n) = -\tfrac{1}{2} \log(1 - \|z\|^2),$$

where $\|z\|^2 = \langle z, z \rangle = \sum_{\ell=1}^{n} |z_\ell|^2$ is the square of the norm of z with respect to the standard Hermitian product on $\mathbb{C}^n$, and introduce the form $\omega = \mathrm{i}\partial\bar{\partial}\phi$. This is a closed real $(1,1)$-form; we will show that it is compatible with the complex structure. We compute

$$\bar{\partial}\phi = \frac{-\bar{\partial}\big(1 - \sum_{\ell=1}^{n} |z_\ell|^2\big)}{2(1 - \|z\|^2)} = \frac{\sum_{\ell=1}^{n} z_\ell \, \mathrm{d}\bar{z}_\ell}{2(1 - \|z\|^2)},$$

which yields

$$\partial\bar{\partial}\phi = \frac{1}{2} \left(\frac{\big(\sum_{k=1}^{n} \bar{z}_k \, \mathrm{d}z_k\big) \wedge \big(\sum_{\ell=1}^{n} z_\ell \, \mathrm{d}\bar{z}_\ell\big)}{(1 - \|z\|^2)^2} + \frac{\sum_{\ell=1}^{n} \mathrm{d}z_\ell \wedge \mathrm{d}\bar{z}_\ell}{1 - \|z\|^2} \right).$$

We finally obtain that

$$\omega = \frac{\mathrm{i}}{2(1 - \|z\|^2)^2} \sum_{k,\ell=1}^{n} \big(\bar{z}_k z_\ell + \big(1 - \|z\|^2\big)\delta_{k,\ell}\big) \, \mathrm{d}z_k \wedge \mathrm{d}\bar{z}_\ell;$$

we claim that the matrix $H = \left(\bar{z}_k z_\ell + \left(1 - \|z\|^2\right)\delta_{k,\ell}\right)_{1 \leq k,\ell \leq n}$ is Hermitian positive definite for every $z \in \mathbb{D}^n$, which means that ω is compatible with the complex structure. Indeed, for a nonzero u in $\mathbb{C}^n$, we have

$$\langle Hu, u \rangle = \langle u, z \rangle \langle z, u \rangle + \left(1 - \|z\|^2\right)\|u\|^2 = |\langle z, u \rangle|^2 + \left(1 - \|z\|^2\right)\|u\|^2 > 0$$

since $1 - \|z\|^2 > 0$.

Example 2.5.9 (The Fubini–Study structure). Let $M = \mathbb{CP}^n$ with its standard open covering $\mathbb{CP}^n \subset \bigcup_{k=0}^n U_k$ where $U_k = \{[z_0 : \cdots : z_n] \in \mathbb{CP}^n \mid z_k \neq 0\}$ and charts

$$\varphi_k : U_k \to \mathbb{C}^n, \quad [z_0 : \cdots : z_n] \mapsto (w_1, \ldots, w_n) = \left(\frac{z_0}{z_k}, \ldots, \frac{z_{k-1}}{z_k}, \frac{z_{k+1}}{z_k}, \ldots, \frac{z_n}{z_k}\right).$$

On each U_k we can define a function

$$\phi_k = \log\left(\sum_{\ell=0}^n \left|\frac{z_\ell}{z_k}\right|^2\right) = \log\left(1 + \sum_{m=1}^n |w_m|^2\right),$$

which, as we will prove, is a local Kähler potential. We define real $(1,1)$-forms ω_k on each U_k by $\omega_k = \mathrm{i}\partial\bar{\partial}\phi_k$. Firstly, we check that this defines a global element $\omega \in \Omega^{1,1}(M) \cap \Omega^2(M)$, i.e. that

$$\omega_{k|U_k \cap U_\ell} = \omega_{\ell|U_k \cap U_\ell};$$

on $U_k \cap U_\ell$, we have

$$\phi_k = \log\left(\left|\frac{z_\ell}{z_k}\right|^2 \sum_{m=0}^n \left|\frac{z_m}{z_\ell}\right|^2\right) = \log\left(\left|\frac{z_\ell}{z_k}\right|^2\right) + \phi_\ell.$$

Hence, we only need to show that $\partial\bar{\partial}\log\left(|z_\ell/z_k|^2\right) = 0$ on $U_k \cap U_\ell$. This follows from the fact that on $\mathbb{C}$

$$\partial\bar{\partial}\log|w|^2 = \partial\left(\frac{w\,\mathrm{d}\overline{w}}{|w|^2}\right) = \partial\left(\frac{\mathrm{d}\overline{w}}{\overline{w}}\right) = 0.$$

Now, a computation similar to the one in the previous example yields

$$\omega_k = \frac{\mathrm{i}}{(1 + \|w\|^2)^2} \sum_{\ell,m=1}^n \left((1 + \|w\|^2)\delta_{\ell,m} - \overline{w}_\ell w_m\right) \mathrm{d}w_\ell \wedge \mathrm{d}\overline{w}_m.$$

Let $H = \left((1 + \|w\|^2)\delta_{\ell,m} - \overline{w}_\ell w_m\right)_{1 \leq \ell,m \leq n}$ and consider $u \neq 0$ in $\mathbb{C}^n$; then

$$\langle Hu, u \rangle = \|u\|^2 + \|w\|^2\|u\|^2 - |\langle w, u \rangle|^2 \geq \|u\|^2 > 0$$

by the Cauchy–Schwarz inequality. Consequently, $\omega_{\mathrm{FS}} = \omega$ is a Kähler form, called the *Fubini–Study* form. Sometimes its definition involves a factor $\pm 1/(2\pi)$, so that the integral of ω_{FS} on $\mathbb{CP}^1 \subset \mathbb{CP}^n$ is equal to ± 1. In our setting, it is better not to include this factor, as we will see later.

2.6 A Few Useful Properties

Let (M, ω, j) be a Kähler manifold and let $g = \omega(\cdot, j\cdot)$ be the induced Kähler metric. The gradient with respect to g of a function f and the Hamiltonian vector field associated with f are related as follows.

Lemma 2.6.1. *Let $f \in \mathcal{C}^1(M)$. Then $\mathrm{grad}_g\, f = -jX_f$.*

Proof. On the one hand, by definition, the gradient of f is such that the equation $\mathrm{d}f = g(\cdot, \mathrm{grad}_g\, f) = \omega(\cdot, j\,\mathrm{grad}_g\, f)$ holds. But on the other hand, the Hamiltonian vector field of f satisfies $\mathrm{d}f + \omega(X_f, \cdot) = 0$.

Like any other Riemannian metric, the Kähler metric g induces a volume form μ_g on M. But the symplectic form ω also defines a volume form, namely the Liouville volume form $\mu = \omega^{\wedge n}/n!$. They are also related.

Lemma 2.6.2. *These two volume forms are equal: $\mu = \mu_g$.*

Proof. Let us use local complex coordinates $(z_1, \ldots, z_n)$ and let us introduce the real local coordinates $(x_1, y_1, \ldots, x_n, y_n)$ satisfying $z_\ell = x_\ell + iy_\ell$ for every $\ell \in [\![1, n]\!]$. Then we can write

$$\omega = i \sum_{\ell,m=1}^n h_{\ell,m}\, \mathrm{d}z_\ell \wedge \mathrm{d}\bar{z}_m$$

for some functions $h_{\ell,m}$ such that $H(p) = \big(h_{\ell,m}(p)\big)$ is a positive definite Hermitian matrix for every p. Consequently,

$$\mu = i^n \det(H)\, \mathrm{d}z_1 \wedge \mathrm{d}\bar{z}_1 \wedge \cdots \wedge \mathrm{d}z_n \wedge \mathrm{d}\bar{z}_n = 2^n \det(H)\, \mathrm{d}x_1 \wedge \mathrm{d}y_1 \wedge \cdots \wedge \mathrm{d}x_n \wedge \mathrm{d}y_n.$$

Note that $2^n \det(H) = \sqrt{\det g}$; this is a consequence of the definition of H, because

$$h_{\ell,m} = \tfrac{1}{4}g(\partial_{x_\ell} - i\partial_{y_\ell}, \partial_{x_m} + i\partial_{y_m}).$$

Therefore, we finally obtain that

$$\mu = \sqrt{\det g}\, \mathrm{d}x_1 \wedge \mathrm{d}y_1 \wedge \cdots \wedge \mathrm{d}x_n \wedge \mathrm{d}y_n = \mu_g,$$

which was to be proved.   □

In what follows, we will also need the following result, which can be derived from the Hodge theory of compact Kähler manifolds. We do not want to spend time on this theory in these notes; therefore we will not give a proof of this result. It is a consequence of [24, Corollary 3.2.10] for example.

Lemma 2.6.3 (The global i$\partial\bar{\partial}$-lemma). *Let (M,ω) be a compact Kähler manifold. Let α be an exact, real-valued form of type $(1,1)$ on M. Then there exists a function $\phi \in \mathcal{C}^\infty(M,\mathbb{R})$ such that $\alpha = i\partial\bar{\partial}\phi$. This function is unique up to the addition of a constant.*

Chapter 3
Complex Line Bundles with Connections

Let us now recall some facts about complex line bundles. A certain number of definitions and properties could be stated for general vector bundles, but we prefer to focus on the one-dimensional case, since this is the case that will be encountered in the following sections.

3.1 Complex Line Bundles

As before, let M be a smooth manifold.

Definition 3.1.1. A *complex line bundle* over M is the data of a smooth manifold L and a smooth map $\pi\colon L \to M$ such that

(1) for every $m \in M$, the fibre $L_m := \pi^{-1}(m)$ over m is a one-dimensional complex vector space,
(2) M is covered by local trivialisations $(U_i, \tau_i)_{i \in I}$, where $U_i \subset M$ is an open set and

$$\tau_i\colon U_i \times \mathbb{C} \to \pi^{-1}(U_i)$$

is a diffeomorphism which, when restricted to $\{m\} \times \mathbb{C}$, $m \in U_i$, induces a linear isomorphism $\tau_i^m = \tau_i(m, \cdot)\colon \mathbb{C} \to L_m$.

In this definition, one can replace the model vector space $\mathbb{C}$ by any other one-dimensional complex vector space. We will use the notation $L \to M$ or $\pi\colon L \to M$ when we want to keep track of the projection.

Example 3.1.2. $L = M \times \mathbb{C} \to M$ with natural projection is a complex line bundle, called the *trivial line bundle*.

A map $s\colon M \to L$ such that $\pi \circ s$ is the identity is called a *section* of $L \to M$. The space of smooth sections of $L \to M$ will be denoted by $\mathcal{C}^\infty(M, L)$. We can also work with local sections, that is sections over open subsets of M; we will use

© Springer International Publishing AG, part of Springer Nature 2018   23
Y. Le Floch, *A Brief Introduction to Berezin–Toeplitz Operators on Compact Kähler Manifolds*, CRM Short Courses,
https://doi.org/10.1007/978-3-319-94682-5_3

the notation $\mathcal{C}^\infty(U, L)$ for the space of smooth sections over some open subset U. Given a local trivialisation (U_i, τ_i), we have a preferred local section, called the *unit section*, given by $s_i = \tau_i(\cdot, 1)$. It is a non-vanishing element of $\mathcal{C}^\infty(U_i, L)$.

We can describe line bundles by gluing local models as follows. Given two trivialisations $(U_i, \tau_i), (U_j, \tau_j)$ such that $U_i \cap U_j \neq \varnothing$, the associated unit sections s_i, s_j satisfy $s_j = f_{ij} s_i$ on $U_i \cap U_j$ for some smooth function $f_{ij} \in \mathcal{C}^\infty(U_i \cap U_j, \mathbb{C} \setminus \{0\})$. Now, if there exists another trivialisation (U_k, τ_k) such that $U_i \cap U_j \cap U_k$ is non-empty, we obtain the equality

$$f_{ik} = f_{ij} f_{jk} \quad \text{on } U_i \cap U_j \cap U_k, \tag{3.1}$$

called the *cocycle relation*.

Example 3.1.3. The transition functions of the trivial bundle $L = M \times \mathbb{C} \to M$ are the constant functions equal to 1.

Proposition 3.1.4. *Let M be a manifold endowed with an open cover $(U_i)_{i \in I}$ such that there exists a collection $(f_{ij})_{i,j \in I, U_i \cap U_j \neq \varnothing}$ of elements of $\mathcal{C}^\infty(U_i \cap U_j, \mathbb{C} \setminus \{0\})$ satisfying the cocycle relation* (3.1). *Then there exists a line bundle $L \to M$ having the functions f_{ij} as transition functions.*

Proof. Let us consider the binary relation $\sim$ defined on the disjoint union $\bigsqcup_{i \in I} U_i \times \mathbb{C}$ by the condition

$$(m_i, z_i) \in U_i \times \mathbb{C} \sim (m_j, z_j) \in U_j \times \mathbb{C} \iff m_i = m_j \text{ and } z_i = f_{ij}(m_i) z_j.$$

Since the functions f_{ij} satisfy the cocycle relation, this relation is reflexive (because $f_{ii} = 1$), symmetric (because $f_{ij} f_{ji} = 1$) and transitive. Hence it is an equivalence relation, we construct L as the quotient of $\bigsqcup_{i \in I} U_i \times \mathbb{C}$ by $\sim$. It is a smooth manifold, and the map

$$\pi \colon L \to M, \quad [(m_i, z_i)] \to m_i$$

is well-defined and smooth. It is clear that the fibre $\pi^{-1}(m)$ at m is a one-dimensional complex vector space. We now define

$$\tau_i \colon U_i \times \mathbb{C} \to \pi^{-1}(U_i), \quad (m_i, z_i) \to [(m_i, z_i)];$$

they are diffeomorphisms by construction, and induce isomorphisms on fibres. Moreover, it follows from the equality

$$[(m_j, 1)] = [(m_j, f_{ij}(m_j))]$$

that the functions f_{ij} are transition functions for L. $\qquad\square$

Two complex line bundles $\pi \colon L \to M, \tilde{\pi} \colon K \to M$ over the same base manifold are said to be *isomorphic* if there exists a diffeomorphism $\phi \colon L \to K$ such that $\tilde{\pi} \circ \phi = \pi$ and which restricts to an isomorphism on each fibre (such a ϕ is called a line bundle isomorphism).

Lemma 3.1.5. *Let $(f_{ij})_{i,j \in I}$ and $(\tilde{f}_{ij})_{i,j \in I}$ be the transition functions (with respect to the same open covering $(U_i)_{i \in I}$ of M) for $\pi \colon L \to M$ and $\tilde{\pi} \colon K \to M$, respectively. Then $L \to M$ and $K \to M$ are isomorphic if and only if there exist nowhere vanishing smooth functions $(g_i)_{i \in I}$ such that $\tilde{f}_{ij} = \frac{g_i}{g_j} f_{ij}$.*

Proof. Assume that there exists a line bundle isomorphism $\phi \colon L \to K$. For $m \in M$, let $\phi_m \colon L_m \to K_m$ be the restriction of ϕ to the fibre L_m. Let $s_i, \tilde{s}_i$ be the unit sections over U_i of L and K, respectively. Since ϕ_m is an isomorphism, it sends $s_i(m)$ to a generator of K_m, and thus, there exists a complex number $g_i(m) \neq 0$ such that $\phi(m)(s_i(m)) = g_i(m)\tilde{s}_i(m)$. The function $g_i \colon U_i \to \mathbb{C} \setminus \{0\}$ thus defined is smooth. Consider now $m \in U_i \cap U_j$. On the one hand, we have that

$$\phi_m\big(s_j(m)\big) = \phi_m\big(f_{ij}(m)s_i(m)\big) = f_{ij}(m)g_i(m)\tilde{s}_i(m).$$

On the other hand,

$$\phi_m\big(s_j(m)\big) = g_j(m)\tilde{s}_j(m) = g_j(m)\tilde{f}_{ij}(m)\tilde{s}_i(m),$$

and comparing the two equalities yields $\tilde{f}_{ij}(m) = \big(g_i(m)/g_j(m)\big)f_{ij}(m)$.

Conversely, let us assume that such functions $(g_i)_{i \in I}$ exist. Let $\tau_i \colon U_i \times \mathbb{C} \to \pi^{-1}(U_i)$ and $\tilde{\tau}_i \colon U_i \times \mathbb{C} \to \tilde{\pi}^{-1}(U_i)$ be the trivialisation diffeomorphisms of L and K, respectively. Define ϕ_i on $\pi^{-1}(U_i)$ as

$$\forall (m, z) \in U_i \times \mathbb{C}, \quad \phi_i\big(\tau_i(m, z)\big) = \big(m, g_i(m)\tilde{\tau}_i^m(z)\big).$$

We need to check that this defines a global map $\phi \colon L \to K$, i.e. that if m belongs to $U_i \cap U_j$ and $\tau_j(m, w) = \tau_i(m, z)$, then $\phi_j\big(\tau_j(m, w)\big) = \phi_i\big(\tau_i(m, z)\big)$. This follows from the equalities

$$g_j(m)\tilde{\tau}_j^m(w) = g_j(m)\tilde{f}_{ij}(m)\tilde{\tau}_i^m(w) = f_{ij}(m)g_i(m)\tilde{\tau}_i^m(w) = g_i(m)\tilde{\tau}_i^m\big(f_{ij}(m)\big)$$

and from the fact that

$$zs_i(m) = \tau_j^m(w) = ws_j(m) = wf_{ij}(m)s_i(m)$$

which implies that $wf_{ij}(m) = z$. The map ϕ thus obtained is clearly a diffeomorphism, and its restriction to the fibre over m is an isomorphism. $\qquad\square$

Let us now assume that M is a complex manifold.

Definition 3.1.6. A complex line bundle $L \to M$ is said to be *holomorphic* if L is a complex manifold, $\pi \colon L \to M$ is a holomorphic map, and the trivialisation maps τ_i are biholomorphisms.

Similarly to smooth sections of a smooth line bundle, it makes sense to talk about holomorphic sections of a holomorphic line bundle. As before, we have unit sections $s_i = \tau_i(\cdot, 1)$ which are holomorphic, and we can construct a holomorphic line bundle over M if and only if there exists an open cover $(U_i)_{i \in I}$ of M and a collection $(f_{ij})_{i,j \in I, U_i \cap U_j \neq \varnothing}$ of holomorphic functions from $U_i \cap U_j$ to $\mathbb{C}^*$ satisfying the cocycle relation (3.1).

Example 3.1.7 (Tautological line bundles over projective spaces). Let $M = \mathbb{CP}^n$ and consider the set

$$\mathcal{O}(-1) = \{([u], v) \in \mathbb{CP}^n \times \mathbb{C}^{n+1} \mid v \in \mathbb{C}u\} \subset \mathbb{CP}^n \times \mathbb{C}^{n+1}$$

and the projection $\pi \colon \mathcal{O}(-1) \to \mathbb{CP}^n$ defined by $\pi([u], v) = [u]$. We claim that $\pi \colon \mathcal{O}(-1) \to \mathbb{CP}^n$ is a holomorphic line bundle over $\mathbb{CP}^n$, called the *tautological line bundle*. Indeed, consider the usual open covering $(U_i)_{0 \le i \le n}$ of $\mathbb{CP}^n$:

$$U_i = \{[z_0 : \cdots : z_n] \in \mathbb{CP}^n \mid z_i \neq 0\},$$

and define functions

$$\tau_i \colon U_i \times \mathbb{C} \to \pi^{-1}(U_i), \quad ([u], z) \mapsto \left([u], \frac{z}{u_i} u\right).$$

This expression does not depend on the choice of the representative of $[u]$. It is obviously smooth and has a smooth inverse $\tau_i^{-1}([u], v) = ([u], v_i)$. Therefore $\mathcal{O}(-1)$ is a complex line bundle with local trivialisations (U_i, τ_i). The transition functions

$$f_{ij} \colon U_i \cap U_j \to \mathbb{C} \setminus \{0\}, \quad [u] \mapsto \frac{u_i}{u_j}$$

are holomorphic.

3.2 Operations on Line Bundles

We can perform a certain number of operations on complex line bundles. In these notes, the reader is assumed to be familiar with vector bundles and operations on them, but we want to enter into details in the specific case of line bundles, since we will need some particular results later. Let M, N be smooth manifolds, and let $\pi \colon L \to N$ be a complex line bundle. If $f \colon M \to N$ is a smooth function, we can define the *pullback bundle* $f^*L \to M$ as

$$f^*L := \{(m, u) \in M \times L \mid f(m) = \pi(u)\} \subset M \times L$$

with projection $\tilde{\pi}(m, u) = m$. One can check that this defines a line bundle over M; indeed, given a local trivialisation (U_i, τ_i) of $L \to N$, the map

$$\zeta_i \colon f^{-1}(U_i) \times \mathbb{C} \to \tilde{\pi}^{-1}\big(f^{-1}(U_i)\big), \quad (m, z) \mapsto \tau_i(f(m), z)$$

is a diffeomorphism which restricts to a linear isomorphism $\mathbb{C} \to \tilde{\pi}^{-1}(m)$ for every $m \in f^{-1}(U_i)$, therefore $(f^{-1}(U_i), \zeta_i)$ is a local trivialisation for f^*L. Any smooth section of $L \to N$ defines a smooth section $f^*s = s \circ f$ of $f^*L \to M$, called the *pullback section* of s.

If L, K are two complex line bundles over the same manifold M, we can define their tensor product $L \otimes K \to M$ whose fibre over $m \in M$ is $L_m \otimes K_m$ and whose transition functions are $h_{ij} = f_{ij}g_{ij}$, where f_{ij} and g_{ij} are the transition functions of L and K, respectively. These functions h_{ij} clearly satisfy the cocycle relation (3.1), and the line bundle they generate has fibre $L_m \otimes K_m$ over $m \in M$. It turns out that the tensor product defines a group law on the set of equivalence classes of line bundles (with respect to the relation of being isomorphic as line bundles). By slightly abusing notation, we denote by L^k the k-th tensor power $L^{\otimes k}$ of the line bundle L. The (class of the) trivial line bundle $M \times \mathbb{C}$ is the identity of this group. On the one hand, the inverse of $L \to M$ is the equivalence class of the line bundle whose transition functions are the inverses of the transition functions of L:

$$\forall i, j \in I, \quad g_{ij} = \frac{1}{f_{ij}}.$$

On the other hand, we can construct the dual bundle $L^* \to M$ whose fibre over m is the dual of L_m, with natural projection $\tilde{\pi}$ and with local trivialisations (U_i, ζ_i) where L has local trivialisations (U_i, τ_i) and

$$\zeta_i \colon U_i \times \mathbb{C}^* \to \tilde{\pi}^{-1}(U_i), \quad \zeta_i(m, z) = \big(m, \big((\tau_i^m)^*\big)^{-1}(z)\big).$$

Here $(\tau_i^m)^*$ is the dual map

$$(\tau_i^m)^* \colon L_m^* \to \mathbb{C}^*, \quad \varphi \mapsto \varphi \circ \tau_i^m.$$

Let h_{ij} be the transition functions for this line bundle L^*. By definition of the dual map, we have that

$$1 = (\tau_j^m)^* \left(\big((\tau_j^m)^*\big)^{-1}(1)\right) = \big((\tau_j^m)^*\big)^{-1}\big(\tau_j^m(1)\big)$$

which yields by using the transition functions

$$1 = f_{ij}(m)h_{ij}(m)\big((\tau_i^m)^*\big)^{-1}\big(\tau_i^m(1)\big) = f_{ij}(m)h_{ij}(m).$$

Thus h_{ij} is the inverse of f_{ij}, and we finally obtain that L^{-1} is isomorphic to L^*. The purely tensorial section $\varphi \otimes s$ of $L^* \otimes L$ is identified with the complex-valued function $\varphi(s)$.

Remark 3.2.1. The tensor product endows the set of equivalence classes of line bundles over M with a group structure. This group is isomorphic to the cohomology group $H^2(M, \mathbb{Z})$ [12, Theorem 2.1.3]; let us describe a sketch of proof of this isomorphism, using the language and some results of sheaf cohomology. The transition functions f_{ij} of a line bundle form a degree one Čech cocycle of the open cover $(U_i)_{i \in I}$ with coefficients in the sheaf $\mathcal{C}^\infty(M, \mathbb{C} \setminus \{0\})$ of smooth non-vanishing complex-valued functions. By Lemma 3.1.5, its cohomology class

$[f_{ij}] \in \check{H}^1\big(M, \mathcal{C}^\infty(M, \mathbb{C} \setminus \{0\})\big)$ determines the equivalence class of the associated line bundle. But $\check{H}^1\big(M, \mathcal{C}^\infty(M, \mathbb{C} \setminus \{0\})\big)$ is isomorphic to $\check{H}^2(M, \mathbb{Z})$; the image of $[f_{ij}]$ is the cohomology class of the degree two Čech cocycle

$$\mu_{ijk} = \frac{1}{2\mathrm{i}\pi}\big(\log(f_{ik}) - \log(f_{ij}) - \log(f_{jk})\big).$$

Furthermore, there is an isomorphism $\varphi \colon \check{H}^2(M, \mathbb{Z}) \to H^2(M, \mathbb{Z})$ (see for instance [11, Theorem 15.8]). The composition of these isomorphisms gives the claimed isomorphism between the group of equivalence classes of complex line bundles and $H^2(M, \mathbb{Z})$; the image $c_1(L) := \varphi(\mu_{ijk})$ is called the *first Chern class* of L.

Using the two constructions above, we can form the *external tensor product* of two line bundles $L \to M$ and $K \to N$ which are not necessarily defined over the same base manifold, as

$$L \boxtimes K := p_1^* L \otimes p_2^* K \to M \times N$$

where $p_1 \colon M \times N \to M$ and $p_2 \colon M \times N \to N$ are the natural projections on each factor.

3.3 Connections on Line Bundles

Let us come back to the case where M is a general smooth manifold.

Definition 3.3.1. A *connection* on a complex line bundle $L \to M$ is a linear map $\nabla \colon \mathcal{C}^\infty(M, L) \to \mathcal{C}^\infty(M, T^*M \otimes L)$ satisfying the Leibniz rule:

$$\nabla(fs) = f\nabla s + \mathrm{d}f \otimes s$$

for every $f \in \mathcal{C}^\infty(M)$ and every $s \in \mathcal{C}^\infty(M, L)$. If X is a smooth vector field on M, the map $\nabla_X \colon \mathcal{C}^\infty(M, L) \to \mathcal{C}^\infty(M, L)$ obtained by contracting ∇ and X is called the *covariant derivative* along X.

Example 3.3.2. A section of the trivial line bundle $M \times \mathbb{C} \to M$ identifies with a complex-valued function on M, and for every one-form α, the formula

$$\nabla f = \mathrm{d}f + f\alpha$$

defines a connection on this line bundle. Any connection is of this form with $\alpha = \nabla 1$ where 1 stands for the function which is equal to 1 everywhere.

A connection is a local operator in the following sense: if U is an open subset of M and s is a section of $L \to M$, then the value of ∇s over U only depends on the value of s over U. Indeed, if $s, t \in \mathcal{C}^\infty(M, L)$ agree over U, consider for $m \in U$

a smooth function χ compactly supported in U and equal to one near m. Then $\chi(s - t) = 0$ on U, therefore

$$0 = \nabla\big(\chi(s - t)\big) = \mathrm{d}\chi \otimes (s - t) + \chi\nabla(s - t) = \mathrm{d}\chi \otimes (s - t) + \chi(\nabla s - \nabla t).$$

Since $\mathrm{d}\chi(m) = 0$, this yields $(\nabla s)(m) = (\nabla t)(m)$. Therefore, the connection ∇ defines a local connection ∇^U such that $\nabla^U(s_{|U}) = (\nabla s)_{|U}$ for every section s. We will keep the notation ∇ for ∇^U.

Let (U_i, τ_i) be a local trivialisation of L, and let $s_i = \tau_i(\cdot, 1) \in \mathcal{C}^\infty(U_i, L)$ be the associated unit section. We set $\beta_i = \nabla s_i / s_i \in \Omega^1(U_i) \otimes \mathbb{C}$. Any other section $t \in \mathcal{C}^\infty(U_i, L)$ is of the form $t = f s_i$ for some smooth function f, hence

$$\nabla t = (\mathrm{d}f + f\beta_i) \otimes s_i.$$

Consequently, the connection can be locally identified with $\mathrm{d} + \beta_i$ acting on functions.

Lemma 3.3.3. *If $U_i \cap U_j \neq \varnothing$, the forms β_i, β_j satisfy on $U_i \cap U_j$ the cocycle relation*

$$\beta_j = \beta_i + \frac{\mathrm{d}f_{ij}}{f_{ij}} \tag{3.2}$$

where the functions f_{ij} are the transition functions of $L \to M$.

Proof. On $U_i \cap U_j$, we have that $s_j = f_{ij}s_i$, thus on the one hand

$$\nabla s_j = (\mathrm{d}f_{ij} + f_{ij}\beta_i) \otimes s_i.$$

But on the other hand, $\nabla s_j = \beta_j \otimes s_j = f_{ij}\beta_j \otimes s_i$, therefore (3.2) holds.

Conversely, let M be a manifold endowed with an open cover $(U_i)_{i \in I}$ such that there exists a collection $(f_{ij})_{i,j \in I, U_i \cap U_j \neq \varnothing}$ of elements of $\mathcal{C}^\infty(U_i \cap U_j, \mathbb{C} \setminus \{0\})$ satisfying the cocycle relations (3.1) and (3.2). Then, thanks to Proposition 3.1.4, there exists a line bundle $L \to M$ having the functions f_{ij} as transition functions. This line bundle is endowed with a connection ∇ such that $\nabla s_i = \beta_i s_i$ where s_i is the unit section associated with U_i.

Proposition 3.3.4. *Let $L \to M$ be a complex line bundle. The space of connections on L is non-empty, and of the form*

$$\{\nabla + \beta \mid \beta \in \Omega^1(M) \otimes \mathbb{C}\}$$

where ∇ is any reference connection on $L \to M$.

Proof. For the existence part, we consider a locally finite open cover $(U_i)_{i \in I}$ of M, together with maps $\tau_i : U_i \times \mathbb{C} \to \pi^{-1}(U_i)$, such that (U_i, τ_i) is a local trivialisation. Let $s_i = \tau_i(\cdot, 1)$ be the associated unit section. Let $(\chi_i)_{i \in I}$ be a partition of unity subordinate to the open cover $(U_i)_{i \in I}$. Let $t \in \mathcal{C}^\infty(M, L)$ and for $i \in I$, let $f_i \in \mathcal{C}^\infty(M)$ be such that $t = f_i s_i$ over U_i. Then we define ∇t as

$$\nabla t = \sum_{i \in I} \chi_i \, \mathrm{d} f_i \otimes s_i.$$

By using the Leibniz rule for the differential of functions, it is easily checked that ∇ is indeed a connection.

Now, let ∇, ∇' be two connections on $L \to M$. Then, by the Leibniz rule, for any $s \in \mathcal{C}^\infty(M, L)$ and any $f \in \mathcal{C}^\infty(M)$, we have that

$$\nabla(fs) - \nabla'(fs) = f(\nabla s - \nabla's).$$

This implies that $\nabla - \nabla'$ is the multiplication by some one-form β. $\qquad\square$

One readily checks that if $L \to M$, $L' \to M$ are two line bundles with respective connections ∇, ∇', then $\nabla \otimes \nabla'$ defined by

$$\forall s \in \mathcal{C}^\infty(M, L), \forall t \in \mathcal{C}^\infty(M, L'), \quad (\nabla \otimes \nabla')(s \otimes t) = (\nabla s) \otimes t + s \otimes (\nabla' t)$$

and extended by linearity is a connection on $L \otimes L' \to M$. In particular, a connection on a line bundle induces connections on its positive integer powers. For negative powers of $L \to M$, we only need to show how a connection on L induces a connection ∇^* on $L^{-1} \simeq L^*$. We can do it by asking that $\nabla^* \otimes \nabla = \mathrm{d}$, the usual differential on the trivial line bundle. This means that for every $\varphi \in \mathcal{C}^\infty(M, L^*)$ and every $s \in \mathcal{C}^\infty(M, L)$,

$$\mathrm{d}\big(\varphi(s)\big) = (\nabla^* \varphi)(s) + \varphi(\nabla s).$$

This uniquely determines ∇^*, and we see on this expression that ∇^* satisfies the Leibniz rule.

Now, let M, N be smooth manifolds, let $L \to N$ be a complex line bundle with connection ∇, and let $f \colon M \to N$ be a smooth function. We can define a connection $f^* \nabla$ on the pullback bundle $f^* L \to M$.

Proposition 3.3.5. *There exists a unique connection $f^* \nabla$ on $f^* L \to M$ such that*

$$(f^* \nabla_X)(f^* s) = f^*(\nabla_{\mathrm{d}f(X)} s)$$

for every $s \in \mathcal{C}^\infty(N, L)$ and every vector field X on M.

Proof. This formula defines the value of $(f^* \nabla)(f^* s)$ for every smooth section s of $L \to N$. Now, let t be any element of $\mathcal{C}^\infty(M, f^* L)$. Given $m \in M$, pick a local non-vanishing section u of $L \to N$ near $f(m)$. Then $f^* u$ is a local non-vanishing section of $f^* L \to M$ near m, hence there exists a smooth function g such that $t = g f^* u$ near m. By the Leibniz rule,

$$(f^* \nabla_X) t = \mathrm{d} g(X) f^* u + g(f^* \nabla)(f^* u) = \mathrm{d} g(X) f^* u + g f^*(\nabla_{\mathrm{d}f(X)} u),$$

which defines $f^* \nabla$ for general sections. Indeed, if we choose another local non-vanishing section v of $L \to N$, then we have that $u = \lambda v$ for some local non-vanishing function λ, and $t = h f^* v$ near m, with $h = g f^* \lambda$. But we have that

$$\mathrm{d}h = f^*\lambda\,\mathrm{d}g + gf^*\,\mathrm{d}\lambda, \quad \nabla_{\mathrm{d}f(X)}v = -\frac{\mathrm{d}\lambda(X)}{\lambda^2}u + \frac{1}{\lambda}\nabla_{\mathrm{d}f(X)}u.$$

Consequently, we obtain that

$$\mathrm{d}h(X)f^*v + hf^*(\nabla_{\mathrm{d}f(X)}v) = \mathrm{d}g(X)f^*u + gf^*(\nabla_{\mathrm{d}f(X)}u),$$

hence the value of $(f^*\nabla_X)t$ does not depend on the choice of local section. □

The two previous constructions show that if $L \to M$ and $K \to N$ are two line bundles, the data of a connection on each of them induces a connection on the line bundle $L \boxtimes K \to M \times N$.

3.4 Curvature of a Connection

Definition 3.4.1. Let ∇ be a connection on a line bundle $L \to M$. The *curvature* of ∇ is the differential form $\mathrm{curv}(\nabla) \in \Omega^2(M) \otimes \mathbb{C}$ defined by the formula

$$\mathrm{curv}(\nabla)(X,Y) = \nabla_X\nabla_Y - \nabla_Y\nabla_X - \nabla_{[X,Y]}$$

for every pair of vector fields $X, Y \in C^\infty(M, TM)$.

It is not immediate that this definition makes sense. However, we have the following lemma.

Lemma 3.4.2. *The curvature* $\mathrm{curv}(\nabla)$ *is indeed a two-form.*

Proof. We start by claiming that given a section $s \in C^\infty(M, L)$, the value of the section

$$t := \nabla_X\nabla_Y s - \nabla_Y\nabla_X s - \nabla_{[X,Y]}s$$

at a point $m \in M$ only depends on the values of X, Y at m. For this we consider an arbitrary function $f \in C^\infty(M, \mathbb{R})$ such that $f(m) = 1$, and we compute the value of the section

$$u := \nabla_{fX}\nabla_Y s - \nabla_Y\nabla_{fX}s - \nabla_{[fX,Y]}s$$

at m. Using the Leibniz rule, we get that

$$\nabla_Y\nabla_{fX}s = \nabla_Y(f\nabla_X s) = f\nabla_Y\nabla_X s + (\mathcal{L}_Y f)\nabla_X s.$$

Moreover, the equality $[fX, Y] = f[X, Y] - (\mathcal{L}_Y f)X$ yields

$$\nabla_{[fX,Y]}s = f\nabla_{[X,Y]}s - (\mathcal{L}_Y f)\nabla_X s,$$

so we finally obtain that $u(m) = t(m)$. Hence, $t(m)$ only depends on the value of X at m. Since the same holds for Y, the claim is proved.

Now we show, keeping the above notation, that t is of the form $F_{X,Y}s$ where $F_{X,Y}$ is some smooth function. This will define $\mathrm{curv}(\nabla)(X,Y) = F_{X,Y}$. Because of the previous result, it is enough to work locally near some point $m \in M$. We take a local non-vanishing section s_0 near m; there exists a smooth function g such that $s = gs_0$ near m. We compute locally

$$t = \nabla_X \big((\mathcal{L}_Y g)s_0 + g\nabla_Y s_0\big) - \nabla_Y \big((\mathcal{L}_X g)s_0 + g\nabla_X s_0\big) - (\mathcal{L}_{[X,Y]}g)s_0 - g\nabla_{[X,Y]}s_0.$$

The first term satisfies

$$\nabla_X\big((\mathcal{L}_Y g)s_0 + g\nabla_Y s_0\big) = (\mathcal{L}_X\mathcal{L}_Y g)s_0 + (\mathcal{L}_Y g)\nabla_X s_0 + (\mathcal{L}_X g)\nabla_Y s_0 + g\nabla_X\nabla_Y s_0$$

and we get a similar expression for the second term:

$$\nabla_Y\big(\mathcal{L}_X g)s_0 + g\nabla_X s_0\big) = (\mathcal{L}_Y\mathcal{L}_X g)s_0 + (\mathcal{L}_X g)\nabla_Y s_0 + (\mathcal{L}_Y g)\nabla_X s_0 + g\nabla_Y\nabla_X s_0.$$

Remembering the definition of $[X, Y]$, we obtain that $t = F_{X,Y}s$ with

$$F_{X,Y} = \frac{\nabla_X\nabla_Y s_0 - \nabla_Y\nabla_X s_0 - \nabla_{[X,Y]}s_0}{s_0}.$$

The fact that $\mathrm{curv}(\nabla)$ is bilinear and antisymmetric is obvious.   □

The curvature can be computed locally thanks to the following result.

Lemma 3.4.3. *If* $\nabla = \mathrm{d} + \beta$ *locally, then* $\mathrm{curv}(\nabla) = \mathrm{d}\beta$. *In particular, the curvature of a connection is a closed form.*

Proof. We have the equality

$$\nabla_X\nabla_Y 1 - \nabla_Y\nabla_X 1 - \nabla_{[X,Y]}1 = \nabla_X\big(\beta(Y)\big) - \nabla_Y\big(\beta(X)\big) - \beta([X,Y])$$

which yields

$$\mathrm{curv}(\nabla)(X,Y) = \mathcal{L}_X\big(\beta(Y)\big) - \mathcal{L}_Y\big(\beta(X)\big) - \beta([X,Y]) = \mathrm{d}\beta(X,Y).$$ □

This gives a method to construct local primitives of the curvature: take a local non-vanishing section s of $L \to M$ and compute $\beta = \nabla s/s$. In some sense, every primitive of the curvature can be obtained in this way.

Lemma 3.4.4. *Let* $\beta \in \Omega^1(M) \otimes \mathbb{C}$ *be a primitive of* $\mathrm{curv}(\nabla)$. *Then for every* $m \in M$, *there exists a local non-vanishing section* s *defined on a neighbourhood of* m *such that* $\nabla s = \beta \otimes s$.

Proof. Let $m \in M$ and take any non-vanishing local section s_0 near m. We have $\mathrm{d}\beta_0 = \mathrm{curv}(\nabla)$ where $\beta_0 = \nabla s_0/s_0$. Since $\beta - \beta_0$ is closed, by restricting the neighbourhood of m if necessary, the Poincaré lemma yields a smooth function f such that $\beta - \beta_0 = \mathrm{d}f$ near m. The local section $s = \exp(f)s_0$ satisfies the desired properties.   □

A straightforward computation shows that if $\nabla' = \nabla + \beta$ for some $\beta \in \Omega^1(M) \otimes$ $\mathbb{C}$, then the curvatures satisfy $\mathrm{curv}(\nabla') = \mathrm{curv}(\nabla) + \mathrm{d}\beta$. Thus, it follows from

Proposition 3.3.4 that the cohomology class of the curvature does not depend on the choice of a connection; we will sometimes call this cohomology class the *curvature class* of L. The following result shows how the curvature behaves with respect to the various operations on line bundles.

Proposition 3.4.5. *The following properties hold.*

(1) *If $L \to M$, $L' \to M$ are two line bundles with respective connections ∇, ∇', then*
$$\mathrm{curv}(\nabla \otimes \nabla') = \mathrm{curv}(\nabla) + \mathrm{curv}(\nabla').$$
(2) *If $L \to M$ is a line bundle with connection ∇, the induced connection ∇^* on $L^{-1} \simeq L^*$ satisfies $\mathrm{curv}(\nabla^*) = -\mathrm{curv}(\nabla)$.*
(3) *If $L \to N$ is a line bundle with connection ∇ and $f \colon M \to N$ is a smooth function, then $\mathrm{curv}(f^*\nabla) = f^*\mathrm{curv}(\nabla)$.*

Proof. It suffices to check the results locally.

(1) Let s, t be local non-vanishing sections of $L \to M$ and $L' \to M$ respectively, defined over the same open set $U \subset M$. Write $\nabla s = \beta \otimes s$ and $\nabla t = \gamma \otimes t$. Then $s \otimes t$ is a non-vanishing section of $L \otimes L' \to M$ over U, and

$$(\nabla \otimes \nabla')(s \otimes t) = (\beta s) \otimes t + s \otimes (\gamma t) = (\beta + \gamma)s \otimes t,$$

therefore $\mathrm{curv}(\nabla \otimes \nabla') = \mathrm{d}\beta + \mathrm{d}\gamma = \mathrm{curv}(\nabla) + \mathrm{curv}(\nabla')$.

(2) Let s be a local non-vanishing section of $L \to M$, with $\nabla s = \beta \otimes s$, and let s^* be the unique section of $L^* \to M$ such that $s^*(s) = 1$ (at a point m, $s(m)$ is a basis of L_m and $s^*(m)$ is the dual basis of L_m^*). Then s^* is a local non-vanishing section and

$$0 = \mathrm{d}\big(s^*(s)\big) = \nabla^*(s^*)(s) + s^*(\nabla s) = \nabla^*(s^*)(s) + \beta s^*(s)$$

so $\nabla^* s^* = -\beta \otimes s^*$, which implies the result.

(3) Let $m \in M$, and let s be a non-vanishing section of $L \to N$ near $f(m)$, such that $\nabla s = \beta \otimes s$. Then f^*s is a non-vanishing section of $f^*L \to M$ near m. Furthermore

$$(f^*\nabla)(f^*s) = (f^*\beta)f^*s,$$

thus $\mathrm{curv}(f^*\nabla) = \mathrm{d}(f^*\beta) = f^*(\mathrm{d}\beta) = f^*\big(\mathrm{curv}(\nabla)\big)$. $\qquad\square$

Definition 3.4.6. A connection is said to be *flat* if its curvature vanishes. A *flat line bundle* is a line bundle endowed with a flat connection. A section of a flat line bundle whose covariant derivative vanishes is called a *flat section*.

Lemma 3.4.4 implies that we can always find local flat sections on a flat line bundle.

3.5 The Chern Connection

Definition 3.5.1. Let M be a smooth manifold. A *Hermitian* line bundle is a complex line bundle $L \to M$ endowed with a Hermitian metric h, that is the data of a Hermitian inner product[1] h_m on each fibre L_m, such that the function $h(s, s)$ is smooth for any smooth section s of $L \to M$.

By the polarisation identity, the last condition implies that $h(s, t)$ is smooth for any two smooth sections s, t of $L \to M$. Note that every complex line bundle $L \to M$ is Hermitian. Indeed, let $(U_i, \tau_i)_{i \in I}$ be a cover of M by local trivialisations and let $(f_i)_{i \in I}$ be a partition of unity subordinate to this cover; given $m \in U_i$, define

$$h_m^i(u, v) = (\tau_i^m)^{-1}(u) \, \overline{(\tau_i^m)^{-1}(v)}$$

for any $u, v \in L_m$, where we recall that τ_i^m is the linear isomorphism from $\mathbb{C}$ to L_m induced by τ_i. Then $h = \sum_{i \in I} f_i h^i$ is a Hermitian metric on L.

Definition 3.5.2. A connection ∇ on a Hermitian line bundle $L \to M$ is said to be *compatible* with the Hermitian structure if

$$\mathrm{d}\big(h(s, t)\big) = h(\nabla s, t) + h(s, \nabla t)$$

for every $s, t \in \mathcal{C}^\infty(M, L)$.

Let us look at what happens locally. Let s be a local section of L over some open subset $U \subset M$ such that $h(s, s) = 1$, and let β be the local one-form such that $\nabla s = \beta \otimes s$. Then, for $f, g \in \mathcal{C}^\infty(U)$, a straightforward computation yields

$$h(\nabla(fs), gs) + h\big(fs, \nabla(gs)\big) = \mathrm{d}\big(h(fs, gs)\big) + \big(\beta + \bar{\beta}\big) h(fs, gs).$$

Consequently, ∇ and h are compatible if and only if all the forms β obtained in this way are purely imaginary, i.e. of the form $\beta = \mathrm{i}\alpha$ with $\alpha \in \Omega^1(U)$.

Definition 3.5.3. A connection ∇ on a holomorphic line bundle $L \to M$ over a complex manifold M is said to be *compatible* with the holomorphic structure if

$$s \text{ is a local holomorphic section of } L \to M \iff \forall Z \in \mathcal{C}^\infty(M, T^{1,0}M), \ \nabla_{\bar{Z}} s = 0.$$

If s is a local non-vanishing holomorphic section over some open subset $U \subset M$ and $\nabla s = \beta \otimes s$, then $\nabla_{\bar{Z}} s = 0$ for every Z of type $(1, 0)$ if and only if the form β belongs to $\Omega^{1,0}(U)$.

Proposition 3.5.4. *Let M be a complex manifold and $L \to M$ be a Hermitian holomorphic line bundle. Then there exists a unique connection ∇ on M which is compatible with both the Hermitian and holomorphic structures. The connection ∇ is called the* Chern connection *of $L \to M$.*

---

[1] Our Hermitian inner products are linear in the left variable and semilinear in the right variable.

Proof. Assume that such a connection ∇ exists. Let $s \in \mathcal{C}^\infty(U, L)$ be a local non-vanishing holomorphic section of L defined on some open set $U \subset M$, and consider the non-vanishing function $H = h(s, s)$. Let $\beta \in \Omega^1(U) \otimes \mathbb{C}$ be such that $\nabla s = \beta \otimes s$; since s is holomorphic and ∇ is compatible with the holomorphic structure, we have that $\beta \in \Omega^{1,0}(U)$. Using the compatibility of ∇ with the Hermitian structure, we also have that
$$\mathrm{d}H = h(\nabla s, s) + h(s, \nabla s) = \beta H + \bar{\beta} H.$$
By identifying the $(1, 0)$ parts, this yields
$$\beta = \frac{\partial H}{H} = \partial(\log H),$$
which means that β, hence ∇, is uniquely determined on U.

To prove the existence part, we follow the same reasoning backwards. The fact that the local connections that we obtain glue together to form a global connection comes from the local uniqueness. $\qquad\square$

Observe that the curvature of the Chern connection of a Hermitian holomorphic line bundle lies in $\Omega^{1,1}(M)$ and is purely imaginary in the sense that it is of the form $i\alpha$ for some $\alpha \in \Omega^2(M)$. Indeed, we know from the proof of the uniqueness of the Chern connection that the local forms $\beta_i \in \Omega^1(U_i) \otimes \mathbb{C}$ such that $\nabla s_i = \beta_i s_i$ are given by $\beta_i = \partial(\log(H_i))$ where $H_i = h(s_i, s_i)$. Therefore,
$$\mathrm{curv}(\nabla)_{|U_i} = \mathrm{d}\beta_i = \bar{\partial}\beta_i = \bar{\partial}\partial(\log H_i),$$
which is a purely imaginary form (see for instance the computation in the proof of Lemma 2.4.4). The following results are partial converses of this fact.

Proposition 3.5.5. *Let (M, ω) be a complex manifold, and let $L \to M$ be a complex line bundle over M, endowed with a connection ∇ such that $\mathrm{curv}(\nabla) \in \Omega^{1,1}(M)$. Then there exists a unique holomorphic structure on $L \to M$ which is compatible with ∇.*

Proof. Let $(U_i)_{i \in I}$ be an open cover of M by trivialisation open sets for $L \to M$, and let s_i be the unit section associated with U_i. Let β_i be the differential form such that $\nabla s_i = \beta_i \otimes s_i$. Let us look for a local non-vanishing section $t_i = f_i s_i$ such that $\nabla_{\bar{Z}} t_i = 0$ whenever Z belongs to $T^{1,0}M$. Since
$$\nabla t_i = (\mathrm{d}f_i + f_i \beta_i) \otimes s_i,$$
this amounts to asking that $\mathrm{d}f_i + f_i \beta_i$ is of type $(1, 0)$, i.e. $\bar{\partial}f_i + f_i \beta_i^{(0,1)} = 0$. We claim that we can find such a function f_i. Indeed, since $\mathrm{d}\beta_i = \mathrm{curv}(\nabla) \in \Omega^{1,1}(M)$, we have that $\bar{\partial}\beta_i^{(0,1)} = 0$. Therefore, by taking a smaller U_i if necessary, the Dolbeault–Grothendieck lemma (Lemma 2.4.3) yields a smooth function g_i such that $\beta_i^{(0,1)} = \bar{\partial}g_i$. Hence, the function $f_i = \exp(-g_i)$ satisfies the required property. So we get such a section t_i; now, if $U_i \cap U_j \neq \varnothing$, let F_{ij} be the smooth function such that $t_j = F_{ij} t_i$ on this intersection. Then for every $Z \in \mathcal{C}^\infty(M, T^{1,0}M)$

$$0 = \nabla_{\overline{Z}} t_j = \left(\mathcal{L}_{\overline{Z}} F_{ij}\right) t_i + F_{ij} \nabla_{\overline{Z}} t_i = \left(\mathcal{L}_{\overline{Z}} F_{ij}\right) t_i,$$

consequently F_{ij} is holomorphic. This means that we have found holomorphic transition functions for $L \to M$. For the uniqueness part, observe that each function f_i is defined up to multiplication by some holomorphic function, which does not change the holomorphic structure on the line bundle. $\qquad\square$

In general, a differential form in $\Omega^{1,1}(M) \cap i\Omega^2(M)$ may not be the curvature of a Chern connection. In the compact Kähler case, however, we have the following result.

Proposition 3.5.6. *Let (M,ω) be a compact Kähler manifold, and let $L \to M$ be a holomorphic line bundle over M, with curvature class $-i\omega$. Then there exists a Hermitian metric on $L \to M$, unique up to a multiplicative constant, such that the Chern connection associated with this metric and the initial holomorphic structure has curvature $-i\omega$.*

Proof. Let h_0 be any Hermitian metric on $L \to M$, and let ∇_0 be the corresponding Chern connection. We look for another Hermitian metric, which will be of the form $h = \exp(f)h_0$ for some real-valued smooth function f on M, with Chern connection ∇. Remembering the proof of the uniqueness of the Chern connection, we have that if s is a local non-vanishing holomorphic section and $H = h(s,s)$, then

$$\nabla s = \partial(\log H) \otimes s = \left(\partial f + \partial\big(\log h_0(s,s)\big)\right) \otimes s,$$

which implies that $\mathrm{curv}(\nabla) = \mathrm{curv}(\nabla_0) + \bar\partial\partial f$. Consequently, we want to find some real-valued f solving the equation

$$-i\omega = \mathrm{curv}(\nabla_0) + \bar\partial\partial f.$$

But $i\,\mathrm{curv}(\nabla_0)$ and ω belong to the same cohomology class, hence $i\,\mathrm{curv}(\nabla_0) = \omega + \alpha$ for some exact form α. Furthermore, α is a real-valued form of type $(1,1)$. So by Lemma 2.6.3, there exists a real-valued function f such that $\alpha = i\partial\bar\partial f$; this function solves the above equation. Moreover, f is unique up to the addition of a constant, so h is unique up to multiplication by a constant. $\qquad\square$

Chapter 4
Geometric Quantisation of Compact Kähler Manifolds

Let (M, ω) be a compact, connected, Kähler manifold. The aim of this chapter is to construct a Hilbert space (or rather a family of Hilbert spaces) which will serve as the state space of quantum mechanics associated with the classical phase space M.

4.1 Prequantum Line Bundles

To perform this construction, we need the manifold to be prequantisable, i.e. that there exists a particular line bundle over it, called a prequantum line bundle.

Definition 4.1.1. A *prequantum line bundle* $(L, \nabla, h) \to M$ is a holomorphic Hermitian line bundle whose Chern connection has curvature $\mathrm{curv}(\nabla) = -\mathrm{i}\omega$.

Such a line bundle does not always exist. The following statement describes the precise obstruction to this existence. In what follows, we say that two line bundles with connection (L, ∇) and $(\tilde{L}, \tilde{\nabla})$ over M are equivalent if there exists a line bundle isomorphism $\phi \colon L \to \tilde{L}$ such that $\tilde{\nabla}(\phi \circ s) = \phi \circ (\nabla s)$ for every $s \in \mathcal{C}^{\infty}(M, L)$.

Proposition 4.1.2. *There exists a prequantum line bundle $L \to M$ if and only if the cohomology class $[\omega/(2\pi)]$ lies in the image of $H^2(M, \mathbb{Z})$ in $H^2(M, \mathbb{R})$. When this is the case, the inequivalent choices are parameterised by $H^1(M, \mathbb{T})$.*

This condition amounts to saying that the integral of ω on each generator of $H_2(M, \mathbb{Z})$ belongs to $2\pi\mathbb{Z}$. In particular, if M is a surface, the integrality condition amounts to saying that the symplectic volume of M is an integer multiple of 2π. If M is a simply connected manifold, then $H^1(M, \mathbb{T}) = \{0\}$, which means that if it satisfies the integrality condition, there is only one choice, up to equivalence, of prequantum line bundle.

Proof. Assume that there exists a prequantum line bundle $L \to M$. Then the first Chern class $c_1(L)$ is an integral cohomology class. But one can show (see for instance [12, Theorem 2.2.14]) that it satisfies

© Springer International Publishing AG, part of Springer Nature 2018
Y. Le Floch, *A Brief Introduction to Berezin–Toeplitz Operators on Compact Kähler Manifolds*, CRM Short Courses,
https://doi.org/10.1007/978-3-319-94682-5_4

$$c_1(L) = \left[\frac{i \operatorname{curv}(\nabla)}{2\pi} \right] = \left[\frac{\omega}{2\pi} \right]$$

and thus $[\omega/(2\pi)]$ lies in the image of $H^2(M, \mathbb{Z})$ in $H^2(M, \mathbb{R})$.

Conversely, assume that the cohomology class of $\omega/(2\pi)$ is integral. Then it is the first Chern class of some complex line bundle $L \to M$. Consider any connection ∇ on $L \to M$; by the aforementioned result, the cohomology classes of $\operatorname{curv}(\nabla)$ and of $-i\omega$ coincide. It follows from Proposition 3.3.4 that we can find another connection $\widetilde{\nabla}$ whose curvature is equal to $-i\omega$. Proposition 3.5.5 then implies that we can find a holomorphic structure on $L \to M$ which is compatible with $\widetilde{\nabla}$. This connection $\widetilde{\nabla}$ may not be a Chern connection, but Proposition 3.5.6 yields a Hermitian structure on $L \to M$ such that the corresponding Chern connection has curvature $-i\omega$.

Finally, if there exists a prequantum line bundle L over M, then for any holomorphic Hermitian line bundle $F \to M$ with flat connection, $L \otimes F \to M$ is again a prequantum line bundle. Conversely, if $L \to M$ and $L' \to M$ are two inequivalent prequantum line bundles, then $F = L^{-1} \otimes L'$ is a holomorphic Hermitian line bundle with flat connection. But the holomorphic Hermitian line bundles with flat connections are classified by $H^1(M, \mathbb{T})$ [12, Theorem 2.2.11]. $\qquad\square$

One must keep in mind that the result that we have taken for granted in the proof, stating that the two definitions of the first Chern class—the one from Čech cohomology, and the one from the curvature of a connection—are equivalent, is not trivial. We will need this result again later.

Example 4.1.3 (Complex projective spaces). Let $M = \mathbb{CP}^n$ be endowed with the Fubini–Study form ω_{FS}. Since $H_2(M, \mathbb{Z})$ is generated by the class of $\mathbb{CP}^1 \subset \mathbb{CP}^n$, there exists a prequantum line bundle if and only if the integral of ω_{FS} on $\mathbb{CP}^1$ is 2π times an integer. Let

$$\iota \colon \mathbb{CP}^1 \to \mathbb{CP}^n, \quad [z_0 : z_1] \mapsto [z_0 : z_1 : 0 : \cdots : 0]$$

be the natural embedding of $\mathbb{CP}^1$ into $\mathbb{CP}^n$. On the open set $U_0 = \{z_0 \neq 0\} \subset \mathbb{CP}^1$, we introduce the coordinate $w = z_1/z_0$ so that

$$\iota^* \omega_{\mathrm{FS}} = \frac{i \, dw \wedge d\overline{w}}{(1 + |w|^2)^2}.$$

In polar coordinates $w = \rho \exp(i\theta)$, we have that $dw \wedge d\overline{w} = -2i\rho \, d\rho \wedge d\theta$, thus

$$\int_{\mathbb{CP}^1} \iota^* \omega_{\mathrm{FS}} = 2\pi \int_0^{+\infty} \frac{2\rho \, d\rho}{(1 + \rho^2)^2} = 2\pi.$$

Therefore, there exists a prequantum line bundle for $(M, \omega_{\mathrm{FS}})$ (and it is unique since $\mathbb{CP}^n$ is simply connected). We will see later what this line bundle is.

We will also need the notion of automorphism of a prequantum line bundle.

Definition 4.1.4. An *automorphism* of the prequantum line bundle $(L, \nabla, h) \to M$ is a diffeomorphism φ of L lifting a diffeomorphism φ_M of M, restricting to linear isomorphisms on the fibres and preserving both the Hermitian structure h and the connection ∇.

The last condition means that $\varphi^*(\nabla s) = \nabla(\varphi^* s)$ for every $s \in \mathcal{C}^\infty(M, L)$, where

$$(\varphi^* s)(m) = \varphi^{-1}\Big(s\big(\varphi_M(m)\big)\Big)$$

for every $m \in M$.

4.2 Quantum Spaces

Let (M, ω) be a compact, connected, Kähler manifold, endowed with a prequantum line bundle $L \to M$. The Hermitian metric on L will be denoted by h. Let us recall that in the rest of these notes, we will use the abusive notation L^k to designate $L^{\otimes k}$, $k \geq 1$. For $k \geq 1$, the Hermitian metric h induces a Hermitian metric h_k on the line bundle L^k by defining

$$h_k(u_1 \otimes \cdots \otimes u_k, v_1 \otimes \cdots \otimes v_k) = \prod_{j=1}^k h(u_j, v_j)$$

for pure tensors and imposing sesquilinearity. It also induces a Hermitian metric h_{-1} on the dual bundle L^* such that

$$(h_{-1})_m(\varphi, \phi) = \frac{\varphi(s)\bar{\phi}(s)}{h_m(s, s)}$$

for every $m \in M$, every $\varphi, \phi \in (L_m)^*$ and every $s \in L_m$. This is equivalent to saying that we apply the previous formula with $L_m \otimes L_m^* \simeq \mathbb{C}$.

We endow the space $\mathcal{C}^\infty(M, L^k)$ of smooth sections of L^k with the Hermitian inner product defined by

$$\forall \phi, \psi \in \mathcal{C}^\infty(M, L^k), \quad \langle \phi, \psi \rangle_k = \int_M h_k(\phi, \psi) \, \mathrm{d}\mu$$

where $\mu = |\omega^n|/n!$ is the Liouville measure on M; the associated norm will be denoted by $\|\cdot\|_k$. We define the quantum space at level k as the space

$$\mathcal{H}_k = H^0(M, L^k)$$

of global holomorphic sections of $L^k \to M$. The notation $H^0(M, L^k)$ comes from Dolbeault cohomology; see the next section for some explanations. Before going further, note that it would not have been very useful to consider the space $H^0(M, L^k)$ for $k < 0$, because of the following lemma.

Proposition 4.2.1. *For $k < 0$, we have that $H^0(M, L^k) = \{0\}$.*

Proof. Let $k < 0$, and assume that there exists a non-identically vanishing global holomorphic section ϕ of $L^k \to M$. Let $H = h_k(\phi, \phi)$; since M is compact and H continuous, H reaches a maximum at a point $m_0 \in M$. Necessarily, $H(m_0) > 0$ and thus $H > 0$ on some open neighbourhood U of m_0. By the discussion contained in the previous chapter, the curvature of the connection ∇^k induced by ∇ on L^k coincides with $\mathrm{d}\beta$ on U, where $\beta = \partial H / H$. On the one hand, since L is a prequantum line bundle, we know that $\mathrm{curv}(\nabla^k) = -\mathrm{i}k\omega$. On the other hand, we have that

$$\mathrm{curv}(\nabla^k) = \mathrm{d}\beta = \frac{\mathrm{d}(\partial H)}{H} - \frac{\mathrm{d}H}{H^2} \wedge \partial H = \frac{\bar{\partial}\partial H}{H} - \frac{\mathrm{d}H}{H^2} \wedge \partial H$$

Since $\mathrm{d}H$ vanishes at m_0, we finally obtain that

$$(\bar{\partial}\partial H)_{m_0} = -\mathrm{i}kH(m_0)\omega_{m_0}.$$

In particular, for every $X \in T_{m_0}M$,

$$(\mathrm{i}\bar{\partial}\partial H)_{m_0}(X, j_{m_0}X) = kH(m_0)\omega_{m_0}(X, j_{m_0}X) < 0$$

since j and ω are compatible, $k < 0$ and $H(m_0) > 0$. But we claim that

$$(\mathrm{i}\bar{\partial}\partial H)_{m_0}(X, j_{m_0}X) = -2\big(\mathrm{Hess}_H(m_0)(X, X) + \mathrm{Hess}_H(m_0)(j_{m_0}X, j_{m_0}X)\big),$$

where $\mathrm{Hess}_H(m_0)$ is the Hessian of H at m_0. Since m_0 is a maximum, this Hessian is a non-positive bilinear form, and we obtain a contradiction.

In order to prove the claim, we use local coordinates $(x_\ell, y_\ell)_{1 \leq \ell \leq n}$ near m_0 such that $j\partial_{x_\ell} = \partial_{y_\ell}$ and $j\partial_{y_\ell} = -\partial_{x_\ell}$, and the associated complex coordinates $(z_\ell)_{1 \leq \ell \leq n}$. Then we have that

$$\bar{\partial}\partial H = -\sum_{\ell,m=0}^{n} \frac{\partial^2 H}{\partial z_\ell \partial \bar{z}_m} \, \mathrm{d}z_\ell \wedge \mathrm{d}\bar{z}_m.$$

Moreover, one readily checks that

$$\frac{\partial^2 H}{\partial z_\ell \partial \bar{z}_m} = \frac{\partial^2 H}{\partial x_\ell \partial x_m} + \mathrm{i}\frac{\partial^2 H}{\partial x_\ell \partial y_m} - \mathrm{i}\frac{\partial^2 H}{\partial y_\ell \partial x_m} + \frac{\partial^2 H}{\partial y_\ell \partial y_m}$$

and that

$$\mathrm{d}z_\ell \wedge \mathrm{d}\bar{z}_m = \mathrm{d}x_\ell \wedge \mathrm{d}x_m + \mathrm{i}\mathrm{d}y_\ell \wedge \mathrm{d}x_m - \mathrm{i}\mathrm{d}x_\ell \wedge \mathrm{d}y_m + \mathrm{d}y_\ell \wedge \mathrm{d}y_m.$$

Therefore, a straightforward computation leads to

$$\bar{\partial}\partial H(\partial_{x_\ell}, j\partial_{x_m}) = 2\mathrm{i}\left(\frac{\partial^2 H}{\partial x_\ell \partial x_m} + \frac{\partial^2 H}{\partial y_\ell \partial y_m}\right).$$

We claim that we obtain similar formulas for $\bar{\partial}\partial H(\partial_{x_\ell}, j\partial_{y_m})$, $\bar{\partial}\partial H(\partial_{y_\ell}, j\partial_{y_m})$ and $\bar{\partial}\partial H(\partial_{y_\ell}, j\partial_{x_m})$. $\qquad\square$

Exercise 4.2.2. Check the remaining cases at the end of the proof above (the last claim of the proof).

As a consequence of the following result, the space $\mathcal{H}_k$ introduced above is finite-dimensional for every $k \geq 1$.

Proposition 4.2.3. *Let* $\pi\colon K \to N$ *be a holomorphic line bundle over a compact complex manifold* N. *Then* $\mathcal{H} = H^0(N, K)$ *is finite-dimensional.*

Proof. We will define a norm on $\mathcal{H}$ and prove that the closed unit ball of $\mathcal{H}$ for this norm is compact, which will imply that this space is finite-dimensional. Let $(U_i)_{1 \leq i \leq m}$ be a finite open cover of M by trivialisation open sets, and let $s_i\colon U_i \to \pi^{-1}(U_i)$ be the associated holomorphic unit section. Let $s \in \mathcal{H}$; for every $i \in [\![1, m]\!]$, there exists a holomorphic function $f_i\colon U_i \to \mathbb{C}$ such that $s = f_i s_i$ on U_i. Now, let $(V_i)_{1 \leq i \leq m}$ be a refinement of $(U_i)_{1 \leq i \leq m}$ such that $K_i := \overline{V}_i \subset U_i$ is compact. We define the norm of s as follows:

$$\|s\| = \sum_{i=1}^{m} \|f_i\|_{L^\infty(K_i)}.$$

Now, let $(s^n)_{n \geq 1}$ be a sequence of elements of the unit ball of $\mathcal{H}$, and let $(f_i^n)_{n \geq 1}$, $1 \leq i \leq m$, be the corresponding sequences of functions from U_i to $\mathbb{C}$. By definition of the norm on $\mathcal{H}$, we have

$$\forall n \geq 1, \forall m \in V_i, \quad |f_i^n(m)| \leq 1.$$

By Montel's theorem for holomorphic functions, there exists a subsequence of $(f_1^n)_{n \geq 1}$ converging uniformly to a holomorphic function $f_1\colon K_1 \to \mathbb{C}$. It follows from a diagonal extraction argument that there exist holomorphic functions $f_i\colon K_i \to \mathbb{C}$, $1 \leq i \leq m$, such that a subsequence of $(f_i^n)_{n \geq 1}$ converges uniformly to f_i on K_i. On intersections $V_i \cap V_j$, we have that

$$\forall n \geq 1, \quad f_j^n g_{ij} = f_i^n$$

where g_{ij} are the transition functions for K. By taking the limit, we see that the functions $f_1, \ldots, f_m$ satisfy the same relation. Therefore we can construct a global section $s \in \mathcal{H}$ such that $s_{|U_i} = f_i s_i$. It follows from the definition of the norm that s belongs to the closed unit ball of $\mathcal{H}$ and that s^n converges to s as n goes to infinity. $\square$

Unfortunately, this tells us nothing about the magnitude of this dimension. However, by using more involved methods, one can prove the following result. This will be the subject of the next section.

Theorem 4.2.4. *The dimension of the Hilbert space* $\mathcal{H}_k$ *satisfies*

$$\dim \mathcal{H}_k = \left(\frac{k}{2\pi}\right)^n \mathrm{vol}(M) + O(k^{n-1})$$

when k goes to infinity, where $2n = \dim M$.

Remark 4.2.5. This equivalent of the dimension is in accordance with the uncertainty principle. Indeed, this principle implies that the minimal volume that a state can occupy in phase space is of order $(2\pi\hbar)^n = (2\pi)^n k^{-n}$. In the limit $k \to +\infty$, the states forming a basis of $\mathcal{H}_k$ should fill the whole phase space, so we expect to have the estimate $(2\pi)^n k^{-n} \dim \mathcal{H}_k \sim \text{vol}(M)$.

Remark 4.2.6. It is sometimes useful to add an auxiliary Hermitian holomorphic bundle $K \to M$ and to define the quantum spaces as the spaces of holomorphic sections of the line bundle $L^k \otimes K$. We will not deal with this case in these notes.

4.3 Computation of the Dimension

This section is devoted to the proof of Theorem 4.2.4. Since the results that we will need are far beyond the reach of this course, we will only state them, without giving any proof. Again, we refer the reader to the standard textbooks already mentioned in the previous chapters.

Let N be a complex manifold of real dimension $2n$, and let $F \to N$ be a holomorphic line bundle. Let

$$\Omega^{p,q}(F) = \mathcal{C}^\infty(\Omega^{p,q}(N) \otimes F)$$

be the space of F-valued forms of type (p, q). There exists a natural operator

$$\bar{\partial}_F \colon \Omega^{p,q}(F) \to \Omega^{p,q+1}(F)$$

satisfying the Leibniz rule and such that $\bar{\partial}_F^2 = 0$. This operator is constructed by using the usual $\bar{\partial}$ operator in local trivialisations, and proving that its action does not depend on the chosen trivialisation. We define

$$H^q(N, F) = \frac{\ker\left(\bar{\partial}_F \colon \Omega^{0,q}(F) \to \Omega^{0,q+1}(F)\right)}{\bar{\partial}_F\left(\Omega^{0,q-1}(F)\right)}$$

and we denote by $h^q(N, F)$ the dimension of this space.

Remark 4.3.1. This definition explains the notation $H^0(M, L^k)$ that we have used for the space of holomorphic sections of $L^k \to M$.

The *Euler characteristic* of $F \to N$ is the quantity

$$\chi(N, F) = \sum_{q=0}^{n} h^q(N, F).$$

The Hirzebruch–Riemann–Roch theorem gives a formula to compute this number. We start by defining some characteristic classes.

Definition 4.3.2. Let $F \to N$ be a complex line bundle over a smooth manifold N. The *Chern character* of F at order j is defined as

$$\mathrm{ch}_j(F) = \frac{c_1(F)^{\wedge j}}{j!}$$

where the first Chern class of F is seen as an element of $H^2(N, \mathbb{R})$, and, given differential forms α, β, we have $[\alpha] \wedge [\beta] = [\alpha \wedge \beta]$.

Theorem 4.3.3 (The Hirzebruch–Riemann–Roch theorem). *Let F be a holomorphic line bundle over a compact complex manifold N. Then*

$$\chi(N, F) = \sum_{j=0}^{n} \int_M \mathrm{ch}_j(F) \wedge \mathrm{td}_{n-j}(N).$$

The cohomology class $\mathrm{td}_j(N)$ is the *Todd class* at order j of the manifold N. We will not define it here but we will simply use the fact that $\mathrm{td}_0(N) = 1$. Coming back to our problem where M is a compact Kähler manifold and $L \to M$ is a prequantum line bundle, if we know how to compute the quantities $h^q(M, L^k)$ for $q \geq 1$, then we will be able to use this formula to compute the dimension of $\mathcal{H}_k$.

Definition 4.3.4. Let N be a compact Kähler manifold and let $F \to N$ be a holomorphic line bundle. We say that F is *positive* if there exists α positive in the sense that

$$-\mathrm{i}\alpha(X, \overline{X}) > 0$$

for every $m \in M$ and every $X \neq 0 \in T_m^{1,0}M$, such that $c_1(F) = [\alpha]$.

The *canonical bundle* of a Kähler manifold N is the line bundle $K_N = \Omega^{n,0}(N)$.

Theorem 4.3.5 (The Kodaira vanishing theorem). *Let N be a compact Kähler manifold and let $F \to N$ be a positive line bundle. Then*

$$H^q(N, K_N \otimes F) = \{0\}$$

whenever $q > 0$.

Now, assume that M is a prequantisable compact Kähler manifold and let $L \to M$ be a prequantum line bundle.

Lemma 4.3.6. *There exists $k_0 \geq 1$ such that for every $k \geq k_0$, $K_M^* \otimes L^k$ is positive.*

Exercise 4.3.7. Prove this lemma. *Hint*: Observe that for $X = Y - \mathrm{i}jY \in T^{1,0}M$, the formula $-\mathrm{i}\omega(X, \overline{X}) = 2g(Y, Y)$ holds, where g is the Kähler metric. Use local orthonormal bases of TM with respect to g to conclude.

Applying the Kodaira vanishing theorem to the line bundle $F = K_M^* \otimes L^k$, and using the previous lemma, we obtain that for k large enough, $H^q(M, L^k) = \{0\}$ whenever $q > 0$. Therefore,

$$\dim \mathcal{H}_k = \chi\big(M, L^k\big),$$

and we can use the Hirzebruch–Riemann–Roch formula to compute the right-hand side. Observe that

$$\mathrm{ch}_j\big(L^k\big) = \left[\frac{1}{j!}\left(\frac{\mathrm{i}}{2\pi}\right)^j \mathrm{curv}(\nabla^k)^{\wedge j}\right] = \frac{k^j}{(2\pi)^j\, j!}\big[\omega^{\wedge j}\big]$$

and remember that $\mathrm{td}_0(M) = 1$. Thus

$$\dim \mathcal{H}_k = \left(\frac{k}{2\pi}\right)^n \int_M \frac{\omega^{\wedge n}}{n!} + O\big(k^{n-1}\big) = \left(\frac{k}{2\pi}\right)^n \mathrm{vol}(M) + O\big(k^{n-1}\big).$$

We will see another derivation of this formula later.

4.4 Examples

Let us now describe a few examples of this construction.

Example 4.4.1 (*A non-compact example: the plane*). We start by reviewing the example of the plane $\mathbb{R}^2$, which does not completely fit in the setting introduced above, because it is not compact. Nevertheless, it is an important example; firstly, it serves to understand the previous constructions in a simple context, and secondly it will be useful when studying the case of the two-dimensional torus.

We equip the plane with coordinates (x, ξ) and its standard symplectic form $\omega = \mathrm{d}\xi \wedge \mathrm{d}x$. We identify it with $\mathbb{C}$ by using the complex coordinate $z = (x - \mathrm{i}\xi)/\sqrt{2}$, so that $\omega = \mathrm{i}\mathrm{d}z \wedge \mathrm{d}\bar{z}$. We consider the following primitive of ω:

$$\alpha = \frac{1}{2}(\xi\,\mathrm{d}x - x\,\mathrm{d}\xi) = \frac{\mathrm{i}}{2}(z\,\mathrm{d}\bar{z} - \bar{z}\,\mathrm{d}z).$$

We endow the trivial bundle $L = \mathbb{R}^2 \times \mathbb{C} \to \mathbb{R}^2$ with the connection $\nabla = \mathrm{d} - \mathrm{i}\alpha$, the curvature of which is equal to $-\mathrm{i}\omega$, and with its standard Hermitian structure. Endowing it with the unique holomorphic structure compatible with both the Hermitian structure and ∇ turns it into a prequantum line bundle. In this case, because of the non-compactness, we define $\mathcal{H}_k$ as the space

$$\mathcal{H}_k = H^0(\mathbb{R}^2, L^k) \cap L^2(M, L^k)$$

of holomorphic sections of L^k that are also square integrable. Let us compute $\mathcal{H}_k$.

The holomorphic tangent bundle $T^{1,0}\mathbb{R}^2$ has as a basis

$$\partial_z = \frac{1}{\sqrt{2}}(\partial_x + \mathrm{i}\partial_\xi),$$

and $\partial_{\bar{z}} = \overline{\partial_z}$ is a basis of $T^{0,1}\mathbb{R}^2$. Furthermore, $(\partial_z, \partial_{\bar{z}})$ is dual to $(\,\mathrm{d}z, \,\mathrm{d}\bar{z})$. Therefore, ψ is a holomorphic section of L if and only if

$$0 = \nabla_{\partial_{\bar{z}}}\psi = \frac{\partial\psi}{\partial\bar{z}} + \frac{z}{2}\psi.$$

A globally non-vanishing solution of this equation is

$$\psi : \mathbb{C} \to \mathbb{C}, \quad z \mapsto \exp\left(-\frac{|z|^2}{2}\right).$$

Consequently, ψ^k belongs to $H^0(\mathbb{R}^2, L^k)$. Since this section never vanishes, any other smooth section of L^k is of the form $f\psi^k$ for some smooth function $f : \mathbb{C} \to \mathbb{C}$, and it is holomorphic if and only if

$$0 = \nabla^k_{\partial_{\bar{z}}}(f\psi^k) = \frac{\partial f}{\partial\bar{z}}\psi^k + f\nabla^k_{\partial_{\bar{z}}}\psi^k = \frac{\partial f}{\partial\bar{z}}\psi^k,$$

that is to say if and only if f is a holomorphic function. Remembering the square integrability condition, we obtain that

$$\mathcal{H}_k = \left\{ f\psi^k \,\middle|\, f : \mathbb{C} \to \mathbb{C} \text{ holomorphic}, \int_{\mathbb{C}} |f(z)|^2 \exp(-k|z|^2)\,|\,\mathrm{d}z \wedge \mathrm{d}\bar{z}| < +\infty \right\}.$$

These spaces are known to be Hilbert spaces and are called Bargmann spaces [2, 3].

Remark 4.4.2. This whole discussion can be generalised to $\mathbb{R}^{2n}$ for $n \geq 1$, or to any finite-dimensional symplectic vector space.

Example 4.4.3 (*Another non-compact example: the unit disc*). Since it does not require too much work after what we just did, let us investigate another non-compact example, namely the unit disc $\mathbb{D} \subset \mathbb{C}$, as in Example 2.5.8. We obtained the Kähler form

$$\omega = \frac{\mathrm{i}\,\mathrm{d}z \wedge \mathrm{d}\bar{z}}{2(1 - |z|^2)^2}.$$

As in the previous example, we endow the trivial line bundle $L = \mathbb{D} \times \mathbb{C} \to \mathbb{D}$ with the connection $\nabla = \mathrm{d} - \mathrm{i}\alpha$, where α is the primitive of ω given by

$$\alpha = \frac{\mathrm{i}z\,\mathrm{d}\bar{z}}{2(1 - |z|^2)},$$

with its standard Hermitian structure, and with the unique holomorphic structure which is compatible with both these structures, so that we obtain a prequantum line bundle. As before, we define $\mathcal{H}_k$ as the space of holomorphic sections of $L^k \to \mathbb{D}$ which are also square integrable. A section ψ of $L \to \mathbb{D}$ is holomorphic if and only if it satisfies the condition

$$0 = \nabla_{\partial_{\bar{z}}}\psi = \frac{\partial\psi}{\partial\bar{z}} + \frac{z}{2(1 - |z|^2)}\psi.$$

The section defined by $\psi(z) = \sqrt{1 - |z|^2}$ is a globally non-vanishing solution of this equation. Therefore, we obtain that

$$\mathcal{H}_k = \left\{ f\psi^k \,\middle|\, f \colon \mathbb{C} \to \mathbb{C} \text{ holomorphic}, \int_{\mathbb{C}} |f(z)|^2 (1 - |z|^2)^{k-2} \,|\,\mathrm{d}z \wedge \mathrm{d}\bar{z}| < +\infty \right\}$$

with inner product

$$\langle f\psi^k, g\psi^k \rangle_k = \int_{\mathbb{D}} f(z)\bar{g}(z)(1 - |z|^2)^{k-2} \,|\,\mathrm{d}z \wedge \mathrm{d}\bar{z}|.$$

Exercise 4.4.4. Check that $(\mathcal{H}_k, \langle\,\cdot\,,\cdot\,\rangle_k)$ is indeed a Hilbert space.

Example 4.4.5 (*Complex projective spaces*). We already saw that the compact Kähler manifold $(M = \mathbb{CP}^n, \omega_{\mathrm{FS}})$ is prequantisable. However, we did not exhibit any prequantum line bundle. Recall the definition of the tautological line bundle:

$$\mathcal{O}(-1) = \{([u], v) \in \mathbb{CP}^n \times \mathbb{C}^{n+1} \mid v \in \mathbb{C}u\} \subset \mathbb{CP}^n \times \mathbb{C}^{n+1},$$

with projection $\pi \colon \mathcal{O}(-1) \to \mathbb{CP}^n$ defined by $\pi([u], v) = [u]$. It is a holomorphic line bundle, and it is endowed with a natural Hermitian structure, which is the one induced by the standard Hermitian structure on the trivial bundle $\mathbb{CP}^n \times \mathbb{C}^{n+1}$, namely:

$$h_{[u]}(v, w) = \sum_{i=1}^{n+1} v_i \overline{w}_i.$$

Let ∇ be the Chern connection corresponding to these structures. We will prove that its curvature is equal to $i\omega_{\mathrm{FS}}$, which will show that the dual bundle $L = \mathcal{O}(1)$ of the tautological bundle, with the induced connection, is a prequantum line bundle for $(M, \omega_{\mathrm{FS}})$ (indeed, recall that the curvature of the induced connection on the dual line bundle is the opposite of the curvature of the original connection). In order to do so, let us recall the proof of Proposition 3.5.4. Let $U_0, \dots, U_n$ be the trivialisation open sets introduced earlier: for every $j \in [\![0, n]\!]$, U_j is defined as $U_j = \{[u_0 : \cdots : u_n] \in \mathbb{CP}^n \mid u_j \neq 0\}$. Let us also recall that we have diffeomorphisms

$$\tau_j \colon U_j \times \mathbb{C} \to \pi^{-1}(U_j), \quad ([u], z) \mapsto \left([u], \frac{z}{u_j}u\right)$$

and the associated unit sections are $s_j([u]) = ([u], (1/u_j)u)$. Then $\nabla s_j = \beta_j \otimes s_j$ with

$$\beta_j = \partial(\log H_j)$$

where $H_j = h(s_j, s_j)$. But we have that

$$\log H_j = \log\left(1 + \sum_{m=1}^{n} |w_m|^2\right) = \phi_j$$

in the coordinates $w = (u_0/u_j, \ldots, u_{j-1}/u_j, u_{j+1}/u_j, \ldots, u_n/u_j)$, with ϕ_j the local Kähler potential introduced in Example 2.5.9. Thus

$$\operatorname{curv}(\nabla) = \mathrm{d}\beta_j = \bar{\partial}\partial(\log H_j) = -\partial\bar{\partial}(\log H_j) = \mathrm{i}\omega_{\mathrm{FS}}.$$

It turns out that there exists a nice description of the space $\mathcal{H}_k = H^0(M, L^k)$. The line bundle L^k is often denoted as $\mathcal{O}(k)$.

Proposition 4.4.6. *There is a canonical isomorphism between the space $\mathcal{H}_k$ and the space $\mathbb{C}_k[z_1, \ldots, z_{n+1}]$ of homogeneous polynomials of degree k in $n+1$ complex variables.*

Proof. Let $C_k^\infty(\mathbb{C}^{n+1} \setminus \{0\}, \mathbb{C})$ be the space of smooth homogeneous functions of degree k from $\mathbb{C}^{n+1} \setminus \{0\}$ to $\mathbb{C}$, and consider the map

$$\Phi \colon C^\infty\big(\mathbb{CP}^n, \mathcal{O}(k)\big) \to C_k^\infty(\mathbb{C}^{n+1} \setminus \{0\}, \mathbb{C})$$

defined as follows: for $s \in C^\infty\big(\mathbb{CP}^n, \mathcal{O}(k)\big)$ and $u \in \mathbb{C}^{n+1} \setminus \{0\}$,

$$\Phi(s)(u) = \langle s([u]), u^{\otimes k}\rangle_{\mathcal{O}(k)_{[u]}, \mathcal{O}(-k)_{[u]}}$$

where $\langle \cdot, \cdot \rangle_{\mathcal{O}(k)_{[u]}, \mathcal{O}(-k)_{[u]}}$ is the duality pairing between $\mathcal{O}(k)_{[u]}$ and $\mathcal{O}(-k)_{[u]}$. This map Φ is obviously linear and injective. It is also surjective; indeed, if f is a smooth homogeneous function of degree k on $\mathbb{C}^{n+1} \setminus \{0\}$, we define a smooth section

$$s([u]) = \big([u], f(u)(u^*)^{\otimes k}\big)$$

of $\mathcal{O}(k) \to \mathbb{CP}^n$. Here u^* is the basis of the dual of the line $\mathbb{C}u$ which is dual to u. This section is well-defined because if we choose another representative $v = \lambda u$ of $[u]$, $\lambda \neq 0$, we will have $v^* = \lambda^{-1}u^*$ and so

$$f(v)(v^*)^{\otimes k} = \lambda^k f(u)\lambda^{-k}(u^*)^{\otimes k} = f(u)(u^*)^{\otimes k}.$$

Clearly $\Phi(s) = f$. Hence Φ is an isomorphism.

We claim that Φ restricts to an isomorphism between $\mathcal{H}_k$ and $\mathbb{C}_k[z_1, \ldots, z_{n+1}]$. Indeed, if s is a holomorphic section of $\mathcal{O}(k) \to \mathbb{CP}^n$, then $\Phi(s)$ is a holomorphic function on $\mathbb{C}^{n+1} \setminus \{0\}$. Hartog's theorem (see e.g. [25, Theorem 2.3.2]) implies that it can be extended to a holomorphic function on $\mathbb{C}^{n+1}$. But a degree k homogeneous holomorphic function on $\mathbb{C}^{n+1}$ is a degree k homogeneous polynomial. Indeed, it is equal to the k homogeneous part in its power series expansion. $\square$

In particular, the dimension of $\mathcal{H}_k$ is equal to $\binom{n+k}{n}$. This can be rewritten as

$$\dim \mathcal{H}_k = \prod_{j=0}^{n-1}\left(1 + \frac{k}{n-j}\right)$$

which is equivalent to $k^n/n!$ when k goes to infinity. The following exercise shows that this is consistent with Theorem 4.2.4.

Exercise 4.4.7. Compute the Liouville volume form associated with the Fubini–Study form, and check that the volume of $\mathbb{CP}^n$ is equal to $(2\pi)^n/n!$.

Example 4.4.8 (Two-dimensional tori). In this example, we first show how to generalise Example 4.4.1 when we consider any two-dimensional symplectic vector space with any linear complex structure, and then we apply this to investigate the case of two-dimensional tori. Let (V, ω_V) be a two-dimensional symplectic vector space equipped with a compatible linear complex structure j. As above, we consider the maps

$$\alpha_x : V \to \mathbb{R}, \quad y \mapsto \tfrac{1}{2}\omega_V(x, y)$$

for $x \in V$, and we endow the trivial line bundle $L_V = V \times \mathbb{C}$ with the connection $\nabla = \mathrm{d} - \mathrm{i}\alpha$, the standard Hermitian structure, and the unique holomorphic structure compatible with these two, making it a prequantum line bundle. Given a lattice $\Lambda \subset V$, we want to quantise the torus $\mathbb{T}^2_\Lambda = V/\Lambda$; this will only be possible if the symplectic volume of the fundamental domain D of Λ is an integer multiple of 2π. For the sake of simplicity, we will assume that this volume is equal to 4π, and we refer the reader to [7, Section 4] for the general case (we will explain later why the following method fails when the volume is equal to 2π). In our case the construction of a prequantum line bundle over $\mathbb{T}^2_\Lambda$ can be achieved as follows.

We want to obtain the prequantum line bundle over $\mathbb{T}^2_\Lambda$ as the quotient of L_V by an action of the lattice Λ. Hence we need to lift the action of Λ to L_V in such a way that all the structures on L_V are preserved by this new action. Recall that a prequantum line bundle automorphism is a line bundle automorphism preserving both the Hermitian structure and the connection.

Lemma 4.4.9. *The group of prequantum line bundle automorphisms of L_V lifting translations identifies with $H = V \times \mathbb{S}^1$ with product*

$$(x, u) \star (y, v) = \Big(x + y, uv \exp\big(\mathrm{i}\alpha_x(y)\big) \Big)$$

for every $(x, u), (y, v) \in H$.

The group H endowed with this product is called the *Heisenberg group*.

Proof. Let G be the group of such automorphisms and let $\varphi \in G$; then it is of the form

$$(z, w) \in V \times \mathbb{C} \mapsto \varphi(z, w) = (z + x, u_x(z)w) \in V \times \mathbb{C}$$

for some $x \in V$ and $u_x : V \to \mathbb{C} \backslash \{0\}$ smooth. Since φ preserves the Hermitian structure, $u_x(z)$ belongs to $\mathbb{S}^1$ for every $z \in V$, so it is of the form $u_x(z) = u \exp\big(\mathrm{i}\theta_x(z)\big)$ for some $u \in \mathbb{S}^1$ and some smooth real-valued function θ_x such that $\theta_x(0) = 0$. Now, let $s : z \mapsto \big(z, f(z)\big)$ be a smooth section of L_V. On the one hand, since

$$(\varphi^* s)(z) = \Big(z, f(z + x)u^{-1} \exp\big(-\mathrm{i}\theta_x(z + x)\big) \Big),$$

we obtain that

$$\left(\nabla(\varphi^* s)\right)(z) = \left(z, u^{-1}\exp\left(-i\theta_x(z+x)\right)\left(\mathrm{d}f(z+x) - if(z+x)\left(\alpha_z + \mathrm{d}\theta_x(z+x)\right)\right)\right).$$

On the other hand, a straightforward computation shows that

$$\left(\varphi^*(\nabla s)\right)(z) = \left(z, u^{-1}\exp\left(-i\theta_x(z+x)\right)\left(\mathrm{d}f(z+x) - if(z+x)\alpha_{z+x}\right)\right).$$

Consequently, φ preserves the connection if and only if for every $z \in V$,

$$\mathrm{d}\theta_x(z+x) = \alpha_{z+x} - \alpha_z = \alpha_x.$$

In other words, $\mathrm{d}(\theta_x - \alpha_x) = 0$, hence $\theta_x = \alpha_x$. Therefore $\varphi = \varphi_{x,u}$ where

$$\varphi_{x,u}(z, w) = \left(z + x, u\exp\left(i\alpha_x(z)\right)w\right) \tag{4.1}$$

for every $(z, w) \in V \times \mathbb{C}$. One readily checks that

$$\varphi_{x,u} \circ \varphi_{y,v} = \varphi_{x+y, uv\exp(i\alpha_x(y))},$$

so the map from H to G sending (x, u) to $\varphi_{x,u}$ is a group isomorphism. $\qquad \square$

Formula (4.1) explicitly gives the action of H on L. Our goal is to see the lattice Λ as a subgroup of the Heisenberg group in order to get an action of Λ on L. There are in fact many different ways to do so.

Lemma 4.4.10. *Let $\chi: \Lambda \to \mathbb{S}^1$ be any group morphism. Then the set $G_\chi = \left\{\left(x, \chi(x)\right) \mid x \in \Lambda\right\}$ is a subgroup of H.*

Proof. Obviously the identity element $(0, 1)$ of H belongs to G_χ and the inverse of an element of G_χ is in G_χ. Let $x, y \in \Lambda$; then

$$\left(x, \chi(x)\right) \star \left(y, \chi(y)\right) = \left(x + y, \chi(x)\chi(y)\exp\left(\frac{i}{2}\omega(x, y)\right)\right).$$

Since the volume of Λ is equal to 4π, $\omega(x, y)$ belongs to $4\pi\mathbb{Z}$, hence

$$\left(x, \chi(x)\right) \star \left(y, \chi(y)\right) = \left(x + y, \chi(x)\chi(y)\right) = \left(x + y, \chi(x + y)\right)$$

belongs to G_χ. $\qquad \square$

We see from the computation in this proof that G_χ is not a subgroup of the Heisenberg group when the volume of Λ is 2π instead of 4π. Of course the corresponding torus can still be quantised in this case but another method has to be used (see for instance [7, Section 4] where cocycles for prequantum line bundles are explicitly given). This construction gives an action of Λ on L, lifting the action on E, which preserves both the connection and the Hermitian structure. Since the translations preserve the complex structure j, this action also preserves the holomorphic structure on L. Therefore we obtain, by taking the quotient, a prequantum line bundle

$L_\chi \to \mathbb{T}^2_\Lambda$. One readily checks that the prequantum line bundles corresponding to distinct morphisms from Λ to $\mathbb{S}^1$ are not equivalent. But such a morphism is characterised by an element of $\mathbb{T}^2 = \mathbb{R}^2/\mathbb{Z}^2$; namely, if (e, f) is a basis of Λ and $\chi(e) = \exp(2i\pi\mu)$, $\chi(f) = \exp(2i\pi\nu)$, then

$$\chi(ae + bf) = \exp\big(2i\pi(a\mu + b\nu)\big) =: \chi_{\mu,\nu}(ae + bf)$$

for every $a, b \in \mathbb{Z}^2$. This is consistent with the fact that the inequivalent choices of prequantum line bundles are parameterised by $H^1(\mathbb{T}^2_\Lambda, \mathbb{T}) = \mathbb{T}^2$ (and this is of course not the result of luck but the manifestation of some general property that we will not describe in these notes).

We now consider H with the new product

$$(x, u) \star_k (y, v) = \Big(x + y, uv \exp\big(ik\alpha_x(y)\big)\Big)$$

Then H acts on L^k where the action is given, in the notation of (4.1), by

$$\varphi_{x,u}(z, w) = \big(z + x, u\exp(ik\alpha_x(z))w\big).$$

This induces an action on sections of this line bundle by the formula

$$\big((x, u) \cdot \psi\big)(z) = u \exp\Big(\frac{ik}{2}\omega(x, z)\Big)\psi(z - x)$$

for $z \in V$. So we choose $(\mu, \nu) \in \mathbb{T}^2$ and thus get an action of Λ on L^k and its sections. The Hilbert space

$$\mathcal{H}^{\mu,\nu}_k = H^0\big(\mathbb{T}^2_\Lambda, L^k_{\chi_{\mu,\nu}}\big)$$

of holomorphic sections of $L^k_{\chi_{\mu,\nu}} \to \mathbb{T}^2_\Lambda$ identifies with the subspace of holomorphic sections of $L^k_V \to V$ which are invariant under the action of Λ through $G_{\chi_{\mu,\nu}}$, with inner product

$$(\phi, \psi) \mapsto \langle \phi, \psi \rangle_k = \int_D \phi\overline{\psi}\,|\omega|$$

where D is any fundamental domain of Λ. In other words, $\mathcal{H}^{\mu,\nu}_k$ consists of holomorphic sections ψ such that $T^*_{x,\mu,\nu}\psi = \psi$ for every $x \in \Lambda$, with

$$(T^*_{x,\mu,\nu}\psi)(z) = \chi_{\mu,\nu}(x)^{-1} \exp\Big(-\frac{ik}{2}\omega(x, z)\Big)\psi(x + z)$$

for any $z \in V$. In order to better understand this space, we start by constructing a non-vanishing holomorphic section of $L^k_V \to V$. In order to do so, we choose the basis (e, f) of Λ such that $\omega(e, f) = 4\pi$, and we introduce the complex number $\tau = a + ib$ where $a, b \in \mathbb{R}$ are such that $f = ae + bje$. One readily checks that $V^{1,0}$ is generated by $Z = e - (1/\overline{\tau})f$, hence if (p, q) are coordinates on $\mathbb{R}^2$ associated with the basis (e, f) (so that $e = \partial_p$ and $f = \partial_q$), then $z = p + \tau q$ is a holomorphic coordinate. In these coordinates, $\omega = 4\pi \, dp \wedge dq$ so

$$\alpha_{(p,q)}(\overline{Z}) = -2\pi\left(q + \frac{p}{\tau}\right)$$

Therefore, a section $t \in \mathcal{C}^\infty(V, L_V)$, that is a function $t\colon V \to \mathbb{C}$, is holomorphic if and only if

$$0 = \nabla_{\overline{Z}} t = \frac{\partial t}{\partial p} - \frac{1}{\tau}\frac{\partial t}{\partial q} + 2i\pi\left(q + \frac{p}{\tau}\right)t.$$

If we look for t of the form $t(p,q) = \exp\bigl(2i\pi g(p,q)\bigr)$, this equation amounts to

$$\frac{\partial g}{\partial p} - \frac{1}{\tau}\frac{\partial g}{\partial q} + q + \frac{p}{\tau} = 0.$$

A straightforward computation shows that the quadratic functions g which are solutions of this equation are of the form

$$g(p,q) = \lambda p^2 + (1 + 2\lambda\tau)pq + \tau(1 + \lambda\tau)q^2$$

for some constant λ. If we choose $\lambda = -1/(2\tau)$, we find $g(q,p) = -1/(2\tau)p^2 + \tau/2q^2$; if j is the standard complex structure, so that $\tau = i$, this yields $t(z) = \exp(-2\pi|z|^2)$ and we recover the section introduced in Example 4.4.1. However, we prefer, in order to simplify the following computations, to take $\lambda = 0$, which means that

$$t(p,q) = \exp\bigl(2i\pi q(p + \tau q)\bigr). \tag{4.2}$$

A straightforward computation shows that for $m, n \in \mathbb{Z}$,

$$(T^*_{me+nf,\mu,\nu}t^k)(p,q) = \exp\Bigl(2i\pi\bigl(k(\tau n^2 + 2n(p + \tau q)) - m\mu - n\nu\bigr)\Bigr)t^k(p,q). \tag{4.3}$$

Any holomorphic section of $L_V^k \to V$ is of the form gt^k where $g\colon V \to \mathbb{C}$ is holomorphic, which means that it satisfies $\partial g/\partial q = \tau \partial g/\partial p$. By the above equation, this section is Λ-invariant if and only if the equality

$$g(p+m, q+n) = \exp\Bigl(-2i\pi\bigl(k(\tau n^2 + 2n(p + \tau q)) - m\mu - n\nu\bigr)\Bigr)g(p,q) \tag{4.4}$$

holds for every $(p,q) \in \mathbb{R}^2$. In particular, $g(p + 1, q) = \exp(2i\pi\mu)g(p,q)$ so the function $(q,p) \mapsto \exp(-2i\pi\mu p)g(p,q)$ is 1-periodic; consequently,

$$g(p,q) = \exp(2i\pi\mu p)\sum_{n\in\mathbb{Z}} g_n(q)\exp(2i\pi np)$$

for some smooth functions $g_n\colon \mathbb{R} \to \mathbb{C}$. The condition that g is holomorphic reads

$$g'_n = 2i\pi\tau(n + \mu)g_n$$

for every $n \in \mathbb{Z}$, so there exists $\rho_n \in \mathbb{C}$ such that $g_n(q) = \rho_n \exp(2\mathrm{i}\pi\tau(n + \mu)q)$. Hence we finally obtain that

$$g(z) = \exp(2\mathrm{i}\pi\mu z) \sum_{n \in \mathbb{Z}} \rho_n \exp(2\mathrm{i}\pi n z) \tag{4.5}$$

where we recall that $z = p + \tau q$. On the one hand, by taking $m = 0$, $n = 1$ in (4.4), we obtain that $g(z + \tau) = \exp\big(2\mathrm{i}\pi(\nu - k(\tau + 2z))\big)\, g(z)$, which yields

$$g(z + \tau) = \exp\big(2\mathrm{i}\pi(\nu - k\tau)\big) \exp(2\mathrm{i}\pi\mu z) \sum_{n \in \mathbb{Z}} \rho_{n+2k} \exp(2\mathrm{i}\pi n z).$$

On the other hand, we have that

$$g(z + \tau) = \exp(2\mathrm{i}\pi\mu\tau) \exp(2\mathrm{i}\pi\mu z) \sum_{n \in \mathbb{Z}} \rho_n \exp(2\mathrm{i}\pi n \tau) \exp(2\mathrm{i}\pi n z).$$

Consequently, the sequence $(\rho_n)_{n \in \mathbb{Z}}$ satisfies

$$\forall n \in \mathbb{Z}, \quad \rho_{n+2k} = \exp\Big(2\mathrm{i}\pi\big((n + \mu + k)\tau - \nu\big)\Big)\rho_n, \tag{4.6}$$

and is thus determined by its terms $\rho_0, \ldots, \rho_{2k-1}$. Note that any choice of such coefficients yields an element of $\mathcal{H}_k^{\mu,\nu}$. Indeed, a straightforward induction shows that the above equation yields

$$\rho_{n+2mk} = \exp\Big(2\mathrm{i}m\pi\big((km + n + \mu)\tau - \nu\big)\Big)\rho_n \tag{4.7}$$

for $m, n \in \mathbb{Z}$. But the series $\sum_{m \in \mathbb{Z}} \exp\big(2\mathrm{i}\pi m k(2z + m\tau)\big)$ is normally convergent on compact sets, since $\Im\tau = 4\pi/\omega(e, je) > 0$, hence its sum defines a holomorphic function, and the associated section is Λ-invariant by construction. The space $\mathcal{H}_k^{\mu,\nu}$ is therefore of dimension $2k$, which is consistent with Theorem 4.2.4. It identifies with the space of theta functions of order k, parameter τ and characteristics μ, ν [36, Section I.3].

4.5 Building More Examples

To conclude this chapter, we indicate one way to construct new examples from the ones introduced in the previous section. Let (M_1, ω_1, j_1) and (M_2, ω_2, j_2) be two compact Kähler manifolds, endowed with prequantum line bundles $(L_1, \nabla_1) \to M_1$ and $(L_2, \nabla_2) \to M_2$, respectively. One readily checks that the product $M_1 \times M_2$ is a compact Kähler manifold, and that the external tensor product $(L_1 \boxtimes L_2, \nabla_1 \otimes \nabla_2) \to M_1 \times M_2$, defined at the end of Section 3.2, is a prequantum line bundle. The following results relates the quantum spaces associated with $M_1 \times M_2$ to the quantum spaces associated with M_1 and M_2.

Proposition 4.5.1. *For every $k \geq 1$, there exists an isomorphism*

$$H^0(M_1 \times M_2, L_1^k \boxtimes L_2^k) \simeq H^0(M_1, L_1^k) \otimes H^0(M_2, L_2^k),$$

whose inverse sends $s \otimes t$ to the section $p_1^* s \otimes p_2^* t$.

Before proving this proposition, let us state an intermediate result which is also useful in its own right. We know from Proposition 4.2.3 that $H^0(M_1, L_1)$ is finite-dimensional; let d_1 be its dimension. For $\mathbf{x} = (x_1, \ldots x_{d_1}) \in M_1^{d_1}$, let

$$\mathrm{ev}_{\mathbf{x}} \colon H^0(M_1, L_1) \to (L_1)_{x_1} \times \cdots \times (L_1)_{x_{d_1}}, \quad s \mapsto (s(x_1), \ldots, s(x_{d_1}))$$

be the joint evaluation map at $\mathbf{x}$.

Lemma 4.5.2. *There exists $\mathbf{x} \in M_1^{d_1}$ such that $\mathrm{ev}_{\mathbf{x}}$ is injective.*

Proof. For $x \in M_1$, let $\mathrm{ev}_x \colon H^0(M_1, L_1) \to (L_1)_x$ be the evaluation map sending the section $s \in H^0(M_1, L_1)$ to $s(x)$. This is a linear form, which is identically vanishing if and only if x belongs to the base locus

$$\mathrm{Bs}(L_1) = \{x \in M_1 \mid \forall s \in H^0(M_1, L_1), s(x) = 0\} = Z(s_1) \cap \cdots \cap Z(s_{d_1}),$$

where $s_1, \ldots, s_{d_1}$ is any basis of $H^0(M_1, L_1)$ and $Z(s)$ is the set of zeros of s. Hence, for $x \in M_1$ not belonging to the base locus, $H_x = \ker \mathrm{ev}_x$ is a hyperplane. We claim that there exists $x_1, \ldots, x_{d_1} \in M_1 \setminus \mathrm{Bs}(L_1)$ such that the hyperplanes $H_{x_1}, \ldots, H_{x_{d_1}}$ are in general position, so that

$$H_{x_1} \cap \cdots \cap H_{x_{d_1}} = \{0\}.$$

This yields that $\mathrm{ev}_{\mathbf{x}}$ is injective, where $\mathbf{x} = (x_1, \ldots, x_{d_1})$.

Exercise 4.5.3. Prove the last claim above. *Hint*: Start by showing that there exists $x, y \in M_1$ such that $H_x \neq H_y$.

Proof of Proposition 4.5.1. We may, and will, assume without loss of generality that $k = 1$. Let

$$\phi \colon H^0(M_1, L_1) \otimes H^0(M_2, L_2) \to H^0(M_1 \times M_2, L_1 \boxtimes L_2)$$

be the map sending $s \otimes t$ to $p_1^* s \otimes p_2^* t$, extended by linearity; ϕ is clearly injective. Now, let $\mathbf{x} = (x_1, \ldots, x_{d_1})$ be as in the previous lemma. Since $\mathrm{ev}_{\mathbf{x}}$ is an injective linear map between two spaces of the same dimension, it is surjective. Hence there exists a basis $s_1, \ldots, s_{d_1}$ of $H^0(M_1, L_1)$ such that for any $\ell \in [\![1, d_1]\!]$, $s_\ell(x_\ell) \neq 0$ and $s_\ell(x_m) = 0$ for any $m \neq \ell$. For $u \in H^0(M_1 \times M_2, L_1 \boxtimes L_2)$, let $v \in H^0(M_1 \times M_2, L_1 \boxtimes L_2)$ be the section defined as

$$v(x, y) = u(x, y) - \sum_{\ell=1}^{d_1} s_\ell(x) \otimes \lambda_\ell(y),$$

where $\lambda_\ell(y)$ is such that $u(x_\ell, y) = s_\ell(x_\ell) \otimes \lambda_\ell(y)$. Then for every $m \in [\![1, d_1]\!]$,

$$v(x_m, y) = u(x_m, y) - s_m(x_m) \otimes \lambda_m(y) = 0;$$

hence, by injectivity of $\mathrm{ev}_{\mathbf{x}}$, $v(x, y) = 0$. This proves that ϕ is surjective. $\square$

Chapter 5
Berezin–Toeplitz Operators

5.1 First Definitions and Properties

As before, let (M, ω) be a prequantizable, compact, connected Kähler manifold, let $L \to M$ be a prequantum line bundle, and let $\mathcal{H}_k$ be the associated Hilbert spaces. Let $L^2(M, L^k)$ be the completion of the space of smooth sections of $L^k \to M$ with respect to the inner product $\langle \cdot, \cdot \rangle_k$ introduced earlier, and let Π_k be the orthogonal projector from $L^2(M, L^k)$ to $\mathcal{H}_k$. This projector is often called the *Szegő projector*.

Definition 5.1.1. Let f in $\mathcal{C}^0(M)$. The *Berezin–Toeplitz operator* associated with f is the operator
$$T_k(f) = \Pi_k f \colon \mathcal{H}_k \to \mathcal{H}_k$$
where f stands for the operator of multiplication by f.

Note that $T_k(1) = \mathrm{Id}_{\mathcal{H}_k}$. We will investigate the properties of such operators. Since the norm of Π_k is smaller than one, a first easy result is the following.

Lemma 5.1.2. *For every* $f \in \mathcal{C}^0(M)$, $\|T_k(f)\| \leq \|f\|_\infty$.

Here $\|T\|$ stands for the operator norm of the operator T. We will show how to obtain a lower bound later. The following result shows what happens for adjoints.

Lemma 5.1.3. *If* $f \in \mathcal{C}^0(M)$, *then*
$$T_k(f)^* = T_k(\bar{f}).$$

In particular, if f is real-valued, then $T_k(f)$ is self-adjoint; if f takes its values in $\mathbb{S}^1$, then $T_k(f)$ is a unitary operator.

Proof. Let $\phi, \psi \in \mathcal{H}_k$. Since Π_k is self-adjoint, we have that
$$\langle \Pi_k(f\phi), \psi \rangle_k = \langle f\phi, \Pi_k \psi \rangle_k = \langle f\phi, \psi \rangle_k.$$

We compute the latter quantity:

© Springer International Publishing AG, part of Springer Nature 2018
Y. Le Floch, *A Brief Introduction to Berezin–Toeplitz Operators on Compact Kähler Manifolds*, CRM Short Courses,
https://doi.org/10.1007/978-3-319-94682-5_5

$$\langle f\phi, \psi\rangle_k = \int_M h_k(f\phi, \psi)\, \mathrm{d}\mu = \int_M f h_k(\phi, \psi)\, \mathrm{d}\mu = \int_M h_k(\phi, \bar{f}\psi)\, \mathrm{d}\mu.$$

Thus, we obtain that $\langle \Pi_k(f\phi), \psi\rangle_k = \langle \phi, \bar{f}\psi\rangle_k$ and the same computation as above shows that this last quantity is equal to $\langle \phi, \Pi_k(\bar{f}\psi)\rangle_k$. This proves the statement of the lemma. $\qquad\square$

Furthermore, the Berezin–Toeplitz quantisation is also a positive quantisation in the following sense.

Lemma 5.1.4. *Let $f \in \mathcal{C}^0(M, \mathbb{R})$ be such that $f(m) \geq 0$ for every $m \in M$. Then $T_k(f)$ is a non-negative operator, in the sense that*

$$\langle T_k(f)\phi, \phi\rangle_k \geq 0$$

for every $\phi \in \mathcal{H}_k$.

Proof. Performing the same computation as in the proof of the previous lemma, we find that

$$\langle \Pi_k(f\phi), \phi\rangle_k = \int_M f h_k(\phi, \phi)\, \mathrm{d}\mu \geq 0. \qquad\square$$

5.2 Norm, Product and Commutator Estimates

The properties of Berezin–Toeplitz operators that we proved so far were easy consequences of their definition. However, some of their properties are harder to grasp. For instance, one may ask whether the composition $T_k(f)T_k(g)$ of two Berezin–Toeplitz operators is still a Berezin–Toeplitz operator, or the same question for the commutator $[T_k(f), T_k(g)]$. One might also want to obtain a lower bound for the operator norm of $T_k(f)$.

Fix some Riemannian metric on M, and for $\ell \in \mathbb{N}$, consider the following norm on the space $\mathcal{C}^\ell(M)$:

$$\|f\|_\ell = \sum_{j=0}^{\ell} \|\nabla_{\mathrm{LC}}^j f\|_\infty,$$

where ∇_{LC} is the Levi-Civita connection. Observe that we have the inequality $\|f\|_\ell \leq \|f\|_m$ if $\ell \leq m$. Now, for any $f, g \in \mathcal{C}^\ell(M)$, we define the quantity

$$\|f, g\|_\ell = \sum_{m=0}^{\ell} \|f\|_m \|g\|_{\ell-m}.$$

We also define, for $p, q \in \mathbb{N}$, $q \geq p$, the quantity

$$\|f,g\|_{p,q} = \sum_{m=p}^{q} \|f\|_m \|g\|_{p+q-m},$$

so that $\|f,g\|_\ell = \|f,g\|_{0,\ell}$. To avoid confusion, we now simply write $\|f\|$ for the maximum norm of f. These norms can be extended to the case of vector fields. If T is an operator, we will write $T = O(k^{-N})\|f,g\|_{p,q}$ if there exists some $k_0 \geq 1$ and $C > 0$ depending neither on f nor on g such that its operator norm is smaller than $Ck^{-N}\|f,g\|_{p,q}$ for $k \geq k_0$. The following precise estimates have been recently obtained in [20].

Theorem 5.2.1. *There exists $C > 0$ such that for every $f \in C^2(M,\mathbb{R})$,*

$$\|T_k(f)\| \geq \|f\| - Ck^{-1}\|f\|_2.$$

This result will be proved in Chapter 9.

Theorem 5.2.2. *For any $f \in C^1(M,\mathbb{R})$ and $g \in C^2(M,\mathbb{R})$,*

$$T_k(f)T_k(g) = T_k(fg) + O(k^{-1})(\|f\|_0\|g\|_2 + \|f\|_1\|g\|_1)$$

and

$$T_k(g)T_k(f) = T_k(fg) + O(k^{-1})(\|f\|_0\|g\|_2 + \|f\|_1\|g\|_1).$$

Finally, the following version of the correspondence principle holds.

Theorem 5.2.3. *For any $f,g \in C^3(M,\mathbb{R})$,*

$$[T_k(f), T_k(g)] = \frac{1}{ik}T_k(\{f,g\}) + O(k^{-2})\|f,g\|_{1,3}.$$

We will prove these two results in Chapter 8. In order to do so, we will show how to derive some properties of the Szegő projector in Chapter 7.

Let us give some explicit examples of Berezin–Toeplitz operators, without proof for the moment. The proofs will be given in Chapter 7, where we investigate the asymptotic behaviour of the projector Π_k.

Example 5.2.4 (*Coordinates on $\mathbb{S}^2$*). The two-dimensional sphere $\mathbb{S}^2$ is diffeomorphic to $\mathbb{CP}^1 = \mathbb{C} \cup \{\infty\}$ via the stereographic projection (from the north pole to the equatorial plane)

$$\pi_N(x_1, x_2, x_3) = \frac{x_1 + ix_2}{1 - x_3}, \qquad \pi_N^{-1}(z) = \frac{1}{1 + |z|^2}(2\,\Re(z), 2\,\Im(z), |z|^2 - 1).$$

In this representation, the complex number z is the holomorphic coordinate on the open set $U_1 = \{[z_0 : z_1] \in \mathbb{CP}^1 \mid z_1 \neq 0\}$ introduced in Example 2.5.9. A straightforward computation shows that the pullback of the Fubini–Study form is $\pi_N^*\omega_{FS} = -\omega_{\mathbb{S}^2}/2$ where $\omega_{\mathbb{S}^2}$ is the standard symplectic form on the sphere viewed as a submanifold of $\mathbb{R}^3$:

$$(\omega_{\mathbb{S}^2})_u(v, w) = \langle u, v \wedge w \rangle_{\mathbb{R}^3}$$

for $u \in \mathbb{S}^2$ and $v, w \in T_u\mathbb{S}^2$. In the angular coordinates (θ, φ) such that

$$(x_1, x_2, x_3) = (\cos\theta\sin\varphi, \sin\theta\sin\varphi, \cos\varphi)$$

we have that $\omega_{\mathbb{S}^2} = -\sin\varphi\, d\theta \wedge d\varphi = d\theta \wedge dx_3$. In the isomorphism between the Hilbert spaces quantizing $\mathbb{CP}^1$ and $\mathbb{C}_k[z_1, z_2]$, this point of view corresponds to considering, for $P \in \mathbb{C}_k[z_1, z_2]$, the polynomial $P(z, 1) \in \mathbb{C}[z]$ of degree at most k. We claim that in this representation

$$T_k(x_1) = \frac{1}{k+2}\left((1-z^2)\frac{d}{dz} + kz\right), \qquad T_k(x_2) = \frac{i}{k+2}\left((1+z^2)\frac{d}{dz} - kz\right)$$

and finally

$$T_k(x_3) = \frac{1}{k+2}\left(2z\frac{d}{dz} - k\,\mathrm{Id}\right).$$

Here we have used the slightly abusive notation $T_k(f)$ for $T_k\big((\pi_N^{-1})^*f\big)$, $f \in C^\infty(\mathbb{S}^2)$. These claims will be proved in Example 7.2.5. The operators $T_k(x_j)$, $j = 1, 2, 3$, correspond to components of spin.

Exercise 5.2.5. From these formulas, check that

$$[T_k(x_1), T_k(x_2)] = \frac{2i}{k+2}T_k(x_3).$$

This is consistent with the theorem about the commutator estimates; indeed, one readily checks that $\{x_1, x_2\} = -2x_3$. We can also check the norm correspondence on, for instance, $T_k(x_3)$. It is clear that the monomials $1, z, \ldots, z^k$ are eigenvectors of $T_k(x_3)$ with eigenvalues $-k/(k+2), (2-k)/(k+2), \ldots, k/(k+2)$. Hence, the norm of $T_k(x_3)$ is equal to the absolute value of the largest diagonal element

$$\frac{k}{k+2} = 1 - \frac{2}{k} + O(k^{-2}).$$

But the maximum norm of x_3 is equal to one.

Remark 5.2.6. We also have that

$$[T_k(x_2), T_k(x_3)] = \frac{2i}{k+2}T_k(x_1), \qquad [T_k(x_3), T_k(x_1)] = \frac{2i}{k+2}T_k(x_2).$$

Example 5.2.7 (*On the torus*).
   We come back to Example 4.4.8 where we explained how to quantise $\mathbb{T}_\Lambda^2 = V/\Lambda$, and keep the same notation. One readily checks that the centre of $(H, \star_k)$ is

$$Z_k(H) := \frac{1}{2k}\Lambda \times \mathbb{S}^1.$$

This subgroup naturally acts on $\mathcal{H}_k^{\mu,\nu}$; indeed, if $\psi \in H^0(V, L_V^k)$ is invariant under the action of Λ and (x, u) belongs to $Z_k(H)$, then for every $y \in \Lambda$,

$$\left(y, \chi_{\mu,\nu}(y)\right) \cdot \left((x, u) \cdot \psi\right) = (x, u) \cdot \left(\left(y, \chi_{\mu,\nu}(y)\right) \cdot \psi\right) = (x, u) \cdot \psi,$$

hence $(x, u) \cdot \psi$ is also invariant under the action of Λ. We let the subgroup $\Lambda/2k \times \{1\}$ of $Z_k(H)$ act on $\mathcal{H}_k^{\mu,\nu}$; for $\lambda \in \Lambda$, we introduce the pullback $T_{\lambda/2k}^*$ of the action of $\lambda/2k$, which is such that for every $\psi \in \mathcal{H}_k^{\mu,\nu}$ and $x \in V$,

$$\left(T_{\lambda/2k}^* \psi\right)(x) = \exp\left(-\frac{\mathrm{i}}{4}\omega(\lambda, x)\right) \psi\left(x + \frac{\lambda}{2k}\right).$$

We also consider the function

$$g_\lambda \colon \mathbb{T}_\Lambda^2 \to \mathbb{R}, \quad x \mapsto \exp\left(-\frac{\mathrm{i}}{2}\omega(\lambda, x)\right).$$

We claim that $T_k(g_\lambda) = T_{\lambda/2k}^* + \mathcal{O}(k^{-1})$. The proof of this claim will be sketched in Example 7.2.10, in which we will also express the action of the operators $T_{e/2k}^*$ and $T_{f/2k}^*$ in a particular orthonormal basis.

5.3 Egorov's Theorem for Hamiltonian Diffeomorphisms

Let $\phi \in \mathrm{Ham}(M, \omega)$ be a Hamiltonian diffeomorphism. Let us recall that if f is a function in $\mathcal{C}^\infty(M)$ generating some Hamiltonian flow ψ^t, then the Hamiltonian flow generated by $f \circ \phi$ is $\phi^{-1} \circ \psi^t \circ \phi$. In physical terms, the classical dynamics of $f \circ \phi$ is conjugated to the one of f; but what about the quantum dynamics? The answer is given by Egorov's theorem: up to an error of order $O(k^{-1})$, $T_k(f \circ \phi)$ is conjugated to $T_k(f)$ by a unitary operator.

Let us explain what is meant by quantum dynamics. Let H in $\mathcal{C}^\infty(\mathbb{R} \times M)$ be a time-dependent Hamiltonian generating ϕ, and for $(t, m) \in \mathbb{R} \times M$, consider $H_t(m) = H(t, m)$. Consider the time-dependent family $\left(T_k(H_t)\right)_{t \in \mathbb{R}}$ of self-adjoint Berezin–Toeplitz operators. Its quantum dynamics is given by the associated Schrödinger equation.

Proposition 5.3.1. *Let $k \geq 1$ be fixed. Given any $u_0 \in \mathcal{H}_k$, the ordinary differential equation*

$$\begin{cases} u'(t) = -\mathrm{i}k T_k(H_t)u(t) \\ u(0) = u_0 \end{cases}$$

has a unique solution $u(\cdot, u_0) \in \mathcal{C}^\infty(\mathbb{R}, \mathcal{H}_k)$. Furthermore, there exists a unitary operator $U_k(t) \colon \mathcal{H}_k \to \mathcal{H}_k$, depending smoothly on t, such that for every u_0 in $\mathcal{H}_k$ and every t in $\mathbb{R}$

$$u(t, u_0) = U_k(t)u_0.$$

Proof. Existence and uniqueness come from the standard result for linear ordinary differential equations. Uniqueness implies that the solution depends linearly on the initial data, and thus yields the existence of $U_k(t)$. In order to prove that $U_k(t)$ is unitary, we consider $V_k(t) = U_k(t)^*U_k(t)$ and compute

$$V_k'(t) = \frac{\mathrm{d}}{\mathrm{d}t}(U_k(t)^*)U_k(t) + U_k(t)^*U_k'(t).$$

Using that $U_k'(t) = -\mathrm{i}kT_k(H_t)U_k(t)$ and taking the adjoint, we get that

$$V_k'(t) = \mathrm{i}k\big(U_k(t)^*T_k(H_t)^*U_k(t) - U_k(t)^*T_k(H_t)U_k(t)\big) = 0.$$

Since $V_k(0) = \mathrm{Id}_{\mathcal{H}_k}$, this implies that $V_k(t) = \mathrm{Id}_{\mathcal{H}_k}$ for all t. Since $\mathcal{H}_k$ is finite-dimensional, this proves that $U_k(t)$ is unitary. $\qquad\square$

The family $\big(U_k(t)\big)_{t\in\mathbb{R}}$ of unitary operators thus defined is the *unitary semigroup* associated with $\big(T_k(H_t)\big)_{t\in\mathbb{R}}$. It is the quantum analogue of the Hamiltonian flow generated by H_t. If $H_t = H$ is time-independent, then $U_k(t) = \exp\big(-\mathrm{i}ktT_k(H)\big)$ where exp is the matrix exponential.

Theorem 5.3.2 (Egorov's theorem). *There exists $k_0 \geq 1$ and $C > 0$ such that for every $k \geq k_0$ and for every $t \in [0,1]$,*

$$\|U_k(t)^*T_k(f)U_k(t) - T_k(f \circ \phi^t)\| \leq Ck^{-1}\int_0^t \|H_s, f\circ\phi^s\|_{1,3}\,\mathrm{d}s.$$

We will need the following result in order to prove this theorem.

Lemma 5.3.3. *Let $(f_t)_{t\in\mathbb{R}}$ be a family of functions of $\mathcal{C}^\infty(M)$ depending smoothly on the parameter t. Then $\big(T_k(f_t)\big)_{t\in\mathbb{R}}$ depends smoothly on t and*

$$\frac{\mathrm{d}}{\mathrm{d}t}T_k(f_t) = T_k\left(\frac{\mathrm{d}}{\mathrm{d}t}f_t\right).$$

Proof. Let s,t in $\mathbb{R}$; by linearity of $g \mapsto T_k(g)$, we get that

$$T_k\left(\frac{f_t - f_s}{t-s}\right) = \frac{1}{t-s}\big(T_k(f_t) - T_k(f_s)\big).$$

Thus we have by the triangle inequality

$$\left\|\frac{\mathrm{d}}{\mathrm{d}t}T_k(f_t) - T_k\left(\frac{\mathrm{d}}{\mathrm{d}t}f_t\right)\right\| \leq \left|\frac{\mathrm{d}}{\mathrm{d}t}T_k(f_t) - \frac{1}{t-s}\big(T_k(f_t) - T_k(f_s)\big)\right|$$
$$+ \left\|T_k\left(\frac{f_t - f_s}{t-s} - \frac{\mathrm{d}}{\mathrm{d}t}f_t\right)\right\|.$$

The first term on the right-hand side of this inequality goes to zero when s goes to t by definition. The remaining term also goes to zero when s goes to t because

$$\left\| T_k \left(\frac{f_t - f_s}{t - s} - \frac{\mathrm{d}}{\mathrm{d}t} f_t \right) \right\| \le \left\| \frac{f_t - f_s}{t - s} - \frac{\mathrm{d}}{\mathrm{d}t} f_t \right\|_\infty .$$

Hence $(\mathrm{d}/\mathrm{d}t) T_k(f_t) = T_k\big((\mathrm{d}/\mathrm{d}t) f_t\big)$. $\qquad\qquad\square$

Proof of Theorem 5.3.2. Consider the Berezin–Toeplitz operator $A_k(t) := T_k(f \circ \phi^t)$ and the operator $B_k(t) := U_k(t) A_k(t) U_k(t)^*$; observe that by Lemma 5.3.3

$$A_k'(t) = T_k \left(\frac{\mathrm{d}}{\mathrm{d}t} (f \circ \phi^t) \right) = T_k(\{H_t, f \circ \phi^t\}).$$

Thus, by Theorem 5.2.3, we have that

$$A_k'(t) = ik[T_k(H_t), A_k(t)] + E_k(t).$$

where $\|E_k(t)\| \le Ck^{-1} \|H_t, f \circ \phi^t\|_{1,3}$ for k greater than some $k_0 \ge 1$ (the same for every t). We compute the time derivative of $B_k(t)$:

$$B_k'(t) = -ik T_k(H_t) U_k(t) A_k(t) U_k(t)^* + U_k(t)^* \big(ik[T_k(H_t), A_k(t)] + E_k(t)\big) U_k(t)$$
$$+ ik U_k(t) A_k(t) U_k(t)^* T_k(H_t),$$

which yields

$$B_k'(t) = U_k(t) E_k(t) U_k(t)^*.$$

Here we have used that $U_k(t)$ and $T_k(H_t)$ commute, which can be proved by differentiating the semigroup relation $U_k(t) U_k(s) = U_k(s) U_k(t)$. Since moreover $B_k(0) = A_k(0) = T_k(f)$, integrating this equation gives

$$B_k(t) = T_k(f) + \int_0^t U_k(s) E_k(s) U_k(s)^* \, \mathrm{d}s$$

for $0 \le t \le 1$, so that for such t,

$$A_k(t) = U_k(t)^* T_k(f) U_k(t) + U_k(t)^* \left(\int_0^t U_k(s) E_k(s) U_k(s)^* \, \mathrm{d}s \right) U_k(t).$$

This implies that for $0 \le t \le 1$ (since $U_k(t)$ is unitary):

$$\|A_k(t) - U_k(t)^* T_k(f) U_k(t)\| \le \int_0^t \|E_k(s)\| \, \mathrm{d}s \le Ck^{-1} \int_0^t \|H_s, f \circ \phi^s\|_{1,3} \, \mathrm{d}s. \qquad \square$$

Example 5.3.4. Let us consider the same data as in Example 5.2.4. The sphere $\mathbb{S}^2$ is equipped with the symplectic form $\omega = -\omega_{\mathbb{S}^2}/2$. Let $H = x_3$; the flow ϕ^t of H at time t is the rotation around the vertical axis with angle $2t$. Therefore, if $f = x_1$, then $f \circ \phi^{\pi/4} = -x_2$. So we would like to compare $U_k^* T_k(x_1) U_k$ with $-T_k(x_2)$, where $U_k = \exp\big(-(ik\pi/4) T_k(x_3)\big)$. In order to do so, we define the operator $A_k(t) = \exp\big(-ikt T_k(x_3)\big) \big(T_k(x_1) + i T_k(x_2)\big) \exp\big(ikt T_k(x_3)\big)$ and compute:

$$A'_k(t) = -ik \exp\big(-iktT_k(x_3)\big)\big([T_k(x_3), T_k(x_1)] + i[T_k(x_3), T_k(x_2)]\big) \exp\big(iktT_k(x_3)\big).$$

Here we have used that $T_k(x_3)$ and $\exp(\pm iktT_k(x_3))$ commute. Using the commutation relations for the operators $T_k(x_j)$, $j = 1, 2, 3$, this reads

$$A'_k(t) = -ik \exp\big(-iktT_k(x_3)\big) \left(\frac{2i}{k+2} T_k(x_2) + \frac{2}{k+2} T_k(x_1)\right) \exp\big(iktT_k(x_3)\big),$$

hence $A'_k(t) = \big(-2ik/(k+2)\big) A_k(t)$. Therefore,

$$A_k(t) = \exp\left(-\frac{2ikt}{k+2}\right) A_k(0) = \exp\left(-\frac{2ikt}{k+2}\right) \big(T_k(x_1) + iT_k(x_2)\big).$$

In particular, we obtain that

$$U_k^* \big(T_k(x_1) + iT_k(x_2)\big) U_k = \exp\left(\frac{ik\pi}{2(k+2)}\right) \big(T_k(x_1) + iT_k(x_2)\big).$$

The identification of the self-adjoint part yields:

$$U_k^* T_k(x_1) U_k = \cos\left(\frac{k\pi}{2(k+2)}\right) T_k(x_1) - \sin\left(\frac{k\pi}{2(k+2)}\right) T_k(x_2).$$

This finally yields that $U_k^* T_k(x_1) U_k = -T_k(x_2) + O\big(k^{-1}\big)$, which is consistent with Egorov's theorem.

Remark 5.3.5. Actually, in this example we could get rid of the remainder by correcting $T_k(x_2)$ by multiplication by the factor $(k+2)/k$. This is not a coincidence; it is linked to the fact that the sphere that we are quantizing is a coadjoint orbit of SU(2), and that the quantisation data is SU(2)-equivariant. This implies that there is an exact version of Egorov's theorem for rotations on the sphere.

Remark 5.3.6. The discussion at the beginning of this section remains valid when ϕ is a general symplectomorphism. There exists a way of quantizing such symplectomorphisms as asymptotically unitary operators, called *Fourier integral operators*, satisfying an analogue of Egorov's theorem. Nevertheless, the construction of such operators is much more involved than the one presented here for Hamiltonian diffeomorphisms, and we will not talk about it in these notes.

Chapter 6
Schwartz Kernels

In this section we give a quick review of the notion of section distributions and Schwartz kernels of operators acting on spaces of sections of vector bundles. A good reference for this material is the classical textbook by Hörmander [26].

6.1 Section Distributions of a Vector Bundle

Let X be a compact, smooth manifold with volume form and let $F \to X$ be a Hermitian vector bundle over X, whose Hermitian form will be denoted by h^F (throughout we will follow the convention that Hermitian forms are linear on the left and anti-linear on the right). We endow the space $\mathcal{C}^\infty(X, F)$ of smooth sections of $F \to X$ with the following inner product:

$$\phi, \psi \in \mathcal{C}^\infty(X, F) \mapsto \langle \phi, \psi \rangle_{L^2(X,F)} = \int_X h_x^F\big(\phi(x), \psi(x)\big)\, \mathrm{d}x,$$

where the integral is performed with respect to the given volume form on X. We define the Hilbert space $L^2(X, F)$ of square integrable sections of $F \to X$ as the completion of $\mathcal{C}^\infty(X, F)$ with respect to this scalar product.

Let us define section distributions on X. We endow the space $\mathcal{C}^\infty(X, F)$ of smooth sections of $F \to X$ with the following topology: we choose a finite cover $(U_i)_{1 \leq i \leq p}$ of X by open sets which are charts for trivialisations of both X and F. For every i, we choose an exhaustion of U_i by compact sets, that is an increasing sequence $(K_n^i)_{n \geq 1}$ of compact sets such that $\bigcup_{n \geq 1} K_n^i = U_i$. Then we consider the countable family $(p_{j,n,i,r})_{j,n \geq 1, 1 \leq i \leq p, 1 \leq r \leq \mathrm{rank}(F)}$ of seminorms defined by

$$p_{j,n,i,r}(\phi) = \max_{K_n^i} |\partial^j \phi_r^i|$$

for every $\phi \in \mathcal{C}^\infty(X, F)$, where, by a slight abuse of notation, ϕ_r^i stands for the r-th coordinate of the image of ϕ in the trivialisation associated with U_i. One can

© Springer International Publishing AG, part of Springer Nature 2018
Y. Le Floch, *A Brief Introduction to Berezin–Toeplitz Operators on Compact Kähler Manifolds*, CRM Short Courses,
https://doi.org/10.1007/978-3-319-94682-5_6

check that the topology induced by this family of seminorms does not depend on the various choices, and that it turns $\mathcal{C}^\infty(X, F)$ into a Fréchet space.

Now, we simply define the space $\mathcal{D}'(X, F)$ of distributions as the topological dual space of $\mathcal{C}^\infty(X, F)$ with respect to this topology, that is to say the space of continuous linear forms on $\mathcal{C}^\infty(X, F)$. The duality pairing will be denoted by $(\,\cdot\,|\,\cdot\,)_{\mathcal{D}'(X,F),\mathcal{C}^\infty(X,F)}$, which means that

$$\forall \alpha \in \mathcal{D}'(X, F), \forall \phi \in \mathcal{C}^\infty(X, F), \quad (\alpha \,|\, \phi)_{\mathcal{D}'(X,F),\mathcal{C}^\infty(X,F)} = \alpha(\phi).$$

Given a square integrable section ψ of $F \to X$, we can view it as a section distribution by putting

$$\forall \phi \in \mathcal{C}^\infty(X, F), \quad (\psi \,|\, \phi)_{\mathcal{D}'(X,F),\mathcal{C}^\infty(X,F)} = \langle \phi, \psi \rangle_{L^2(X,F)} = \int_X h^F_x\big(\phi(x), \psi(x)\big)\,\mathrm{d}x;$$

the map $\psi \in L^2(X, F) \mapsto (\psi \,|\, \cdot\,)_{\mathcal{D}'(X,F),\mathcal{C}^\infty(X,F)}$ is injective and this way we get the following inclusions:

$$\mathcal{C}^\infty(X, F) \subset L^2(X, F) \subset \mathcal{D}'(X, F).$$

6.2 The Schwartz Kernel Theorem

Now, let Y be another compact, smooth manifold with volume form and $G \to Y$ be a Hermitian vector bundle over Y. Let

$$p_X : X \times Y \to X, \quad p_Y : X \times Y \to Y$$

be the natural projections. Recall that the external tensor product $F \boxtimes G \to X \times Y$ is defined as

$$F \boxtimes G = p_X^* F \otimes p_Y^* G.$$

Given two sections $\phi \in \mathcal{C}^\infty(X, F)$ and $\psi \in \mathcal{C}^\infty(Y, G)$, we define a section

$$\phi \boxtimes \psi := p_X^* \phi \otimes p_Y^* \psi \in \mathcal{C}^\infty(X \times Y, F \boxtimes G);$$

by a slight abuse of vocabulary, we will call such a section a pure tensor. Furthermore, remember that we can define an Hermitian form $h^{F \boxtimes G}$ on $F \boxtimes G$ by specifying its values on pure tensors as follows:

$$\forall (x, y) \in X \times Y, \forall f \in F_x, \forall g \in G_y, \quad h^{F \boxtimes G}_{(x,y)}(f \otimes g) := h^F_x(f) h^G_y(g),$$

then extending it to the whole $\mathcal{C}^\infty(X \times Y, F \boxtimes G)$ by forcing sesquilinearity.

If V is a complex vector space, let $\overline{V}$ stand for its complex conjugate vector space:

$$\overline{V} := \{\bar{v} \,|\, v \in V\}$$

with addition $\bar{v} + \bar{w} = \overline{v+w}$ and scalar multiplication $\lambda\bar{v} = \overline{\bar{\lambda}v}$. Moreover, if V is equipped with an Hermitian form h^V, we define the Hermitian form $h^{\overline{V}}$ on $\overline{V}$ by the formula

$$\forall \bar{v}, \bar{w} \in \overline{V}, \quad h^{\overline{V}}(\bar{v}, \bar{w}) := \overline{h^V(v, w)}.$$

These definitions extend to complex vector bundles. If $F \to X$ is a complex line bundle with connection ∇, there is an induced connection $\overline{\nabla}$ on $\overline{F} \to X$ such that

$$\overline{\nabla}_X \bar{s} = \overline{\nabla_X s}$$

for every $s \in \mathcal{C}^\infty(X, F)$.

Given a section $K \in \mathcal{C}^\infty(X \times Y, F \boxtimes \overline{G})$, the formula

$$\forall x \in X, \quad (\mathcal{K}\phi)(x) = \int_Y K(x, y) \cdot \phi(y)\, \mathrm{d}y \tag{6.1}$$

defines an operator $\mathcal{K}$ from $\mathcal{C}^\infty(Y, G)$ to $\mathcal{C}^\infty(X, F)$. Here, the dot stands for contraction with respect to the Hermitian product h^G on G:

$$\forall y \in Y, \forall f \in G_y, \forall \bar{g} \in \overline{G}_y, \quad \bar{g} \cdot f := h_y^G(f, g).$$

There exists a generalisation of this construction when K only belongs to the space $\mathcal{D}'(X \times Y, F \boxtimes \overline{G})$. Of course the previous integral would not make sense anymore, but we can build on the following observation.

Lemma 6.2.1. *If $K \in \mathcal{C}^\infty(X \times Y, F \boxtimes \overline{G})$ and $\phi \in \mathcal{C}^\infty(Y, G)$, then*

$$(\mathcal{K}(\phi) \,|\, \psi)_{\mathcal{D}'(X,F),\mathcal{C}^\infty(X,F)} = \left(K \,\middle|\, \psi \boxtimes \bar{\phi}\right)_{\mathcal{D}'(X \times Y, F \boxtimes \overline{G}),\mathcal{C}^\infty(X \times Y, F \boxtimes \overline{G})} \tag{6.2}$$

for every $\psi \in \mathcal{C}^\infty(X, F)$. In this formula we view K as an element of $\mathcal{D}'(X \times Y, F \boxtimes \overline{G})$ and $\mathcal{K}(\phi)$ as an element of $\mathcal{D}'(X, F)$.

Proof. Since the two sides of the equality that we wish to prove are linear with respect to K, it is enough to show that it holds when K is of the form

$$K(x, y) = f(x, y)K_X(x) \boxtimes \overline{K_Y}(y)$$

with $K_X \in \mathcal{C}^\infty(X, F)$, $K_Y \in \mathcal{C}^\infty(Y, G)$. Indeed, we can then conclude by a partition of unity argument. By construction of the injection of $\mathcal{C}^\infty(X, F)$ in $\mathcal{D}'(X, F)$, we have that

$$(\mathcal{K}(\phi) \,|\, \psi)_{\mathcal{D}'(X,F),\mathcal{C}^\infty(X,F)} = \int_X h_x^F(\psi(x), \mathcal{K}(\phi)(x))\, \mathrm{d}x.$$

But we have by definition

$$\mathcal{K}(\phi)(x) = \int_Y K(x, y) \cdot \phi(y)\, \mathrm{d}y = \left(\int_Y f(x, y)h_y^G(\phi(y), K_Y(y))\, \mathrm{d}y\right)K_X(x)$$

and hence, after substituting in the previous equation:

$$\left(\mathcal{K}(\phi) \mid \psi\right)_{\mathcal{D}'(X,F),\mathcal{C}^\infty(X,F)}$$

$$= \int_{X \times Y} \overline{f(x,y)} h_x^F\big(\psi(x), K_X(x)\big) \overline{h_y^G\big(\phi(y), K_Y(y)\big)} \, dx \, dy.$$

But we can write $\overline{h_y^G\big(\phi(y), K_Y(y)\big)} = h_y^{\overline{G}}\big(\bar{\phi}(y), \overline{K_Y}(y)\big)$, so

$$\left(\mathcal{K}(\phi) \mid \psi\right)_{\mathcal{D}'(X,F),\mathcal{C}^\infty(X,F)}$$

$$= \int_{X \times Y} h_{(x,y)}^{F \boxtimes \overline{G}}\big((\psi \boxtimes \bar{\phi})(x,y), f(x,y)\big(K_X \boxtimes \overline{K_Y}\big)(x,y)\big) \, dx \, dy,$$

which is precisely equal to $\big(K \mid \psi \boxtimes \bar{\phi}\big)_{\mathcal{D}'(X \times Y, F \boxtimes \overline{G}), \mathcal{C}^\infty(X \times Y, F \boxtimes \overline{G})}$. $\square$

The advantage of this formulation is that we can now take it as a definition of $\mathcal{K}(\phi)$ when K is not necessarily smooth anymore; (6.2) defines the action of the distribution $\mathcal{K}(\phi)$ on pure tensors and we can then pass to any element of $\mathcal{C}^\infty(X \times Y, F \boxtimes \overline{G})$ by linearity.

Theorem 6.2.2 (The Schwartz kernel theorem). *Every $K \in \mathcal{D}'\big(X \times Y, F \boxtimes \overline{G}\big)$ defines according to (6.2) an operator $\mathcal{K} \colon \mathcal{C}^\infty(Y, G) \to \mathcal{D}'(X, F)$ which is continuous in the sense that for every sequence $(\phi_n)_{n \geq 1}$ of elements of $\mathcal{C}^\infty(Y, G)$ converging to zero, the sequence $\big(\mathcal{K}(\phi_n)\big)_{n \geq 1}$ converges to zero in $\mathcal{D}'(X, F)$. Conversely, any continuous operator $\mathcal{K} \colon \mathcal{C}^\infty(Y, G) \to \mathcal{D}'(X, F)$ has a unique section distribution $K \in \mathcal{D}'\big(X \times Y, F \boxtimes \overline{G}\big)$ such that (6.2) holds; K is called the* Schwartz kernel *of $\mathcal{K}$.*

We will take this theorem for granted. For a proof, one can look at the proof of [26, Theorem 5.2.1] and adapt it to the setting of section distributions.

Example 6.2.3. Assume that $Y = X$ and $G = F$ (endowed with the same Hermitian form). The Schwartz kernel K of the identity $\mathcal{C}^\infty(X, F) \to \mathcal{C}^\infty(X, F)$ is given by the formula

$$\left(K \mid \Psi\right)_{\mathcal{D}'(X \times Y, F \boxtimes \overline{G}), \mathcal{C}^\infty(X \times Y, F \boxtimes \overline{G})} = \int_X \mathrm{Tr}\big(\Psi(x, x)\big) \, dx$$

for every $\Psi \in \mathcal{C}^\infty(X \times X, F \boxtimes \overline{F})$. Here we view $\Psi(x, x) \in F_x \otimes \overline{F}_x$ as an endomorphism of F_x by using the identifications $\overline{F}_x \simeq F_x^*$ and $F_x \otimes F_x^* \simeq \mathrm{End}(F_x)$.

6.3 Operators Acting on Square Integrable Sections

Let X, Y, F, G be as in the previous parts, and consider orthonormal bases $(\psi_i)_{i \geq 1}$ and $(\phi_j)_{j \geq 1}$ of the Hilbert spaces $L^2(X, F)$ and $L^2(Y, G)$, respectively. Then one can check that the sequence $(\psi_i \boxtimes \bar{\phi}_j)_{i,j \geq 1}$ is an orthonormal basis of the space $L^2\big(X \times Y, F \boxtimes \overline{G}\big)$. Given $K(\cdot, \cdot) \in L^2\big(X \times Y, F \boxtimes \overline{G}\big)$, formula (6.1) defines an operator $\mathcal{K}$ from $L^2(Y,G)$ to $L^2(X,F)$ which is continuous in the usual L^2 sense (exercise: check it). The following result is a derivation of the Schwartz kernel theorem in this particular case.

Proposition 6.3.1. *Let $\mathcal{K}\colon L^2(Y,G) \to L^2(X,F)$ be a bounded operator. Then $\mathcal{K}$ has a Schwartz kernel $K(\cdot,\cdot) \in L^2\big(X \times Y, F \boxtimes \overline{G}\big)$, which can be computed as follows. For any two orthonormal bases $(\psi_i)_{i\geq 1}$ of $L^2(X,F)$ and $(\phi_j)_{j\geq 1}$ of $L^2(Y,G)$, we have that*

$$K(x,y) = \sum_{i,j\geq 1} \langle \mathcal{K}\phi_j, \psi_i \rangle_{L^2(X,F)} \psi_i(x) \otimes \overline{\phi_j(y)} \tag{6.3}$$

in the L^2 sense.

Proof. Let $\varphi \in L^2(Y,G)$. Then

$$\mathcal{K}\varphi = \sum_{i\geq 1} \langle \mathcal{K}\varphi, \psi_i \rangle_{L^2(X,F)} \psi_i = \sum_{i,j\geq 1} \langle \varphi, \phi_j \rangle_{L^2(Y,G)} \langle \mathcal{K}\phi_j, \psi_i \rangle_{L^2(X,F)} \psi_i.$$

Observe that for $j \geq 1$, $x \in X$ and $y \in Y$,

$$\big(\overline{\phi_j(y)} \cdot \varphi(y)\big)\psi_i(x) = \big(\psi_i(x) \otimes \overline{\phi_j(y)}\big) \cdot \varphi(y);$$

therefore we have that for every $x \in X$,

$$\langle \varphi, \phi_j \rangle_{L^2(Y,G)} \psi_i(x) = \int_Y \big(\psi_i(x) \otimes \overline{\phi_j(y)}\big) \cdot \varphi(y)\,\mathrm{d}y.$$

Using this in the first equation, we finally obtain that

$$(\mathcal{K}\varphi)(x) = \int_Y \left(\sum_{i,j\geq 1} \langle \mathcal{K}\phi_j, \psi_i \rangle_{L^2(X,F)} \psi_i(x) \otimes \overline{\phi_j(y)} \right) \cdot \varphi(y)\,\mathrm{d}y$$

for every $x \in X$, which was to be proved. $\qquad\square$

For orthogonal projectors, there exists another nice formula, which can be derived either from the previous result or by a direct computation.

Lemma 6.3.2. *Let S be a closed subspace of $L^2(X,F)$ and let Π be the orthogonal projector from $L^2(X,F)$ into S. Let $(\varphi_\ell)_{\ell\geq 1}$ be an orthonormal basis of S. Then the Schwartz kernel K of Π is given by the formula:*

$$K(x,y) = \sum_{\ell\geq 1} \varphi_\ell(x) \otimes \overline{\varphi_\ell}(y).$$

Proof. Let $\phi \in L^2(X,F)$. Using the expression

$$\Pi\phi = \sum_{\ell\geq 1} \langle \phi, \varphi_\ell \rangle_{L^2(X,F)} \varphi_\ell$$

and performing the same computation as in the proof of the proposition above, namely showing that

$$\langle \phi, \varphi_\ell \rangle_{L^2(X,F)} \varphi_\ell(x) = \int_X \left(\varphi_\ell(x) \otimes \overline{\varphi_\ell(y)} \right) \cdot \phi(y) \, \mathrm{d}y,$$

we obtain the result.   □

In what follows, we will often use the abusive notation $K(\,\cdot\,,\cdot\,)$ for the Schwartz kernel of an operator K.

Proposition 6.3.3 (Restriction to subspaces). *Let $K\colon L^2(X,F) \to L^2(X,F)$ be a bounded operator with Schwartz kernel $K(\,\cdot\,,\cdot\,) \in L^2\big(X^2, F \boxtimes \overline{F}\big)$. Moreover, let $S \subset L^2(X,F)$ be a closed subspace of $L^2(X,F)$, and let Π be the orthogonal projector from $L^2(X,F)$ to S. Then the operator $K\Pi$ possesses a Schwartz kernel $\widetilde{K}$ which satisfies, for any orthonormal basis $(\varphi_\ell)_{\ell \geq 1}$ of S,*

$$\widetilde{K}(x,y) = \sum_{\ell \geq 1} (K\varphi_\ell)(x) \otimes \overline{\varphi_\ell(y)}.$$

Proof. Let $\varphi \in L^2(M, L^k)$. Since $\Pi\varphi = \sum_{\ell \geq 1} \langle \phi, \varphi_\ell \rangle_{L^2(X,F)} \varphi_\ell$ and K is bounded,

$$K\Pi\varphi = \sum_{\ell \geq 1} \langle \phi, \varphi_\ell \rangle_{L^2(X,F)} K\varphi_\ell.$$

We use the same computation as in the two previous proofs to finish the proof. □

Proposition 6.3.4 (Trace of an integral operator). *Let $K\colon L^2(X,F) \to L^2(X,F)$ be a bounded operator with Schwartz kernel $K(\,\cdot\,,\cdot\,) \in L^2\big(X^2, F \boxtimes \overline{F}\big)$. Then*

$$\mathrm{Tr}(K) = \int_X \mathrm{Tr}\big(K(x,x)\big) \, \mathrm{d}x.$$

where we see $K(x,x) \in F_x \otimes \overline{F}_x$ as an endomorphism of F_x.

Proof. By (6.3), we have that for $x \in X$,

$$\mathrm{Tr}\big(K(x,x)\big) = \sum_{i,j \geq 1} \langle K\phi_i, \phi_j \rangle_{L^2(X,F)} \mathrm{Tr}\big(\phi_j(x) \otimes \overline{\phi_i(x)}\big).$$

But the endomorphism $\phi_j(x) \otimes \overline{\phi_i(x)}$ of F_x corresponds to $u \in F_x \mapsto h_x\big(u, \phi_i(x)\big)\phi_j(x)$, hence its trace is equal to $h_x\big(\phi_j(x), \phi_i(x)\big)$, thus

$$\mathrm{Tr}\big(K(x,x)\big) = \sum_{i,j \geq 1} \langle K\phi_i, \phi_j \rangle_{L^2(X,F)} h_x\big(\phi_j(x), \phi_i(x)\big).$$

By integrating, we get

$$\int_X \mathrm{Tr}\big(K(x,x)\big)\,\mathrm{d}x = \sum_{i,j\geq 1} \langle K\phi_i, \phi_j\rangle_{L^2(X,F)} \langle \phi_j(x), \phi_i(x)\rangle_{L^2(X,F)},$$

which yields

$$\int_X \mathrm{Tr}\big(K(x,x)\big)\,\mathrm{d}x = \sum_{i\geq 1} \langle K\phi_i, \phi_i\rangle'_{L^2(X,F)} = \mathrm{Tr}(K). \qquad \square$$

Proposition 6.3.5 (Composition and Schwartz kernels). *Let* $K, J \colon C^0(X,F) \to C^0(X,F)$ *be two operators with Schwartz kernels* $K(\cdot,\cdot), J(\cdot,\cdot) \in C^0\big(X^2, F \boxtimes \overline{F}\big)$. *Then their composition* KJ *possesses the Schwartz kernel* $I(\cdot,\cdot)$ *given by*

$$I(x,y) = \int_X K(x,z)\cdot J(z,y)\ \mathrm{d}z$$

where the dot stands for contraction with respect to h.

Proof. For $\varphi \in C^0(X,F)$, we have that

$$K(J\varphi)(x) = \int_X K(x,z)\cdot (J\varphi)(z)\,\mathrm{d}z = \int_X K(x,z)\cdot \left(\int_X J(z,y)\cdot\varphi(y)\,\mathrm{d}y\right)\mathrm{d}z$$

and changing the order of integration we obtain

$$K(J\varphi)(x) = \int_X \left(\int_X K(x,z)\cdot J(z,y)\,\mathrm{d}z\right)\cdot \varphi(y)\,\mathrm{d}y,$$

which was to be proved. $\qquad \square$

Proposition 6.3.6 (Schwartz kernel of the adjoint). *Let* $K \colon L^2(Y,G) \to L^2(X,F)$ *be a bounded operator with Schwartz kernel* $K(\cdot,\cdot) \in L^2\big(X \times Y, F \boxtimes \overline{G}\big)$. *Then the Schwartz kernel* $\widetilde{K} \in L^2\big(Y \times X, G \boxtimes \overline{F}\big)$ *of its adjoint satisfies* $\widetilde{K}(y,x) = \overline{K(x,y)}$.

Proof. Let $\phi \in L^2(Y,G)$ and $\psi \in L^2(X,G)$. Then

$$\langle K\phi, \psi\rangle_{L^2(X,G)} = \int_X h_x^F\left(\int_Y K(x,y)\cdot\phi(y)\,\mathrm{d}y, \psi(x)\right)\mathrm{d}x.$$

Since we have that $h_x^F\big(\int_Y K(x,y)\cdot\phi(y)\,\mathrm{d}y, \psi(x)\big) = \overline{\psi(x)}\cdot\int_Y K(x,y)\cdot\phi(y)\,\mathrm{d}y$, we can rewrite this as

$$\langle K\phi, \psi\rangle_{L^2(X,G)} = \int_Y \left(\int_X \overline{\psi(x)}\cdot K(x,y)\,\mathrm{d}x\right)\cdot\phi(y)\,\mathrm{d}y.$$

But we also have that $\big(\int_X \overline{\psi(x)}\cdot K(x,y)\,\mathrm{d}x\big)\cdot\phi(y) = h_y^G\big(\phi(y), \overline{\int_X \overline{\psi(x)}\cdot K(x,y)\,\mathrm{d}x}\big)$, therefore, since

$$\overline{\psi(x) \cdot K(x,y)} = \overline{h_x^F\big(K(x,y), \psi(x)\big)} = h_x^F\big(\psi(x), K(x,y)\big) = \overline{K(x,y)} \cdot \psi(x),$$

we finally obtain the equality

$$\langle K\phi, \psi \rangle_{L^2(X,G)} = \int_Y h_y^G\left(\phi(y), \int_X \overline{K(x,y)} \cdot \psi(x)\,\mathrm{d}x\right) \mathrm{d}y,$$

which concludes the proof. $\square$

6.4 Further Properties

Norm Estimate

The following result provides with a method to derive estimates for the norm of an operator from estimates for its Schwartz kernel.

Proposition 6.4.1 (The Schur test). *Let T be an endomorphism of $\mathcal{C}^0(X, F)$ with Schwartz kernel $K \in \mathcal{C}^0(X^2, F \boxtimes \overline{F})$. Define the two quantities*

$$C_1 = \sup_{x \in X} \int_X \|K(x,y)\|\,\mathrm{d}y, \quad C_2 = \sup_{y \in X} \int_X \|K(x,y)\|\,\mathrm{d}x.$$

Then the operator norm of T satisfies $\|T\|^2 \le C_1 C_2$.

In the course of the proof of this result, we will need the following lemma.

Lemma 6.4.2. *Let $x \in X$, $G \in \mathcal{C}^0(X^2, F \boxtimes \overline{F})$ and $\varphi \in \mathcal{C}^0(X, F)$. Then for every $f \in L^2(X)$,*

$$\left\| \int_X f(y) G(x,y) \cdot \varphi(y)\,\mathrm{d}y \right\|^2 \le \left(\int_X |f(y)|^2 \|G(x,y)\|^2\,\mathrm{d}y \right)\left(\int_X \|\varphi(y)\|^2\,\mathrm{d}y \right).$$

Proof. Let d be the rank of F, and let $\alpha_1, \ldots, \alpha_d$ be an orthonormal basis of F_x. For every $i \in [\![1, d]\!]$, we define a section $G_x^i \in \mathcal{C}^0(X, F)$ by the formula

$$G_x^i(y) = \alpha_i \cdot \overline{G(x,y)}.$$

Let $y \in X$ and let $\beta_1, \ldots, \beta_d$ be an orthonormal basis of F_y. Then $(\alpha_i \otimes \overline{\beta}_j)_{1 \le i,j \le d}$ is an orthonormal basis of $F_x \otimes \overline{F}_y$, and we write $G(x,y) = \sum_{i,j=1}^d \lambda_{ij} \alpha_i \otimes \overline{\beta}_j$ for some complex numbers λ_{ij}. Then $G_x^i(y) = \sum_{j=1}^d \overline{\lambda}_{ij} \beta_j$, so $\|G_x^i(y)\|^2 = \sum_{j=1}^d |\lambda_{ij}|^2$. Therefore

$$\sum_{i=1}^d \|G_x^i(y)\|^2 = \|G(x,y)\|^2. \tag{6.4}$$

Let us also write $\varphi(y) = \sum_{\ell=1}^{d} \mu_\ell \beta_\ell$. On the one hand, $G(x, y) \cdot \varphi(y) = \sum_{i,j=1}^{d} \lambda_{ij} \mu_j \alpha_i$. But on the other hand, $h_y(\varphi(y), G_x^i(y)) = \sum_{j=1}^{d} \lambda_{ij} \mu_j$. Consequently, we obtain the following formula:

$$G(x, y) \cdot \varphi(y) = \sum_{i=1}^{d} h_y(\varphi(y), G_x^i(y)) \, \alpha_i$$

We deduce from this equality that

$$\int_X f(y) G(x, y) \cdot \varphi(y) \, \mathrm{d}y = \sum_{i=1}^{d} \langle \varphi, \bar{f} G_x^i \rangle_{L^2(X, F)} \alpha_i.$$

Thus, we have that

$$\left\| \int_X f(y) G(x, y) \cdot \varphi(y) \, \mathrm{d}y \right\|^2 = \sum_{i=1}^{d} |\langle \varphi, \bar{f} G_x^i \rangle_{L^2(X, F)}|^2$$
$$\leq \left(\sum_{i=1}^{d} \|\bar{f} G_x^i\|_{L^2(X, F)}^2 \right) \|\varphi\|_{L^2(X, F)}^2,$$

where the last equality follows from the Cauchy–Schwarz inequality. But (6.4) implies that

$$\sum_{i=1}^{d} \|\bar{f} G_x^i\|_{L^2(X, F)}^2 = \sum_{i=1}^{d} \int_X |f(y)|^2 \|G_x^i(y)\|^2 \, \mathrm{d}y = \int_X |f(y)|^2 \|G(x, y)\|^2 \, \mathrm{d}y,$$

which concludes the proof. $\qquad \square$

Proof of Proposition 6.4.1. Let $\varphi \in \mathcal{C}^0(X, F)$ and let $x \in X$. Let $U_x \subset X$ be the open subset consisting of the points $y \in X$ such that $K(x, y) \neq 0$. Then

$$(T\phi)(x) = \int_X 1_{U_x}(y) \frac{K(x, y)}{\sqrt{\|K(x, y)\|}} \cdot \sqrt{\|K(x, y)\|} \, \phi(y) \, \mathrm{d}y$$

where 1_{U_x} is the indicator function of U_x. Applying the previous lemma with $f = 1_{U_x}$, $G(x, y) = K(x, y)/\sqrt{\|K(x, y)\|}$ and $\varphi(y) = \sqrt{\|K(x, y)\|} \, \phi(y)$, we get

$$\|(T\phi)(x)\|^2 \leq \left(\int_X \|K(x, y)\| \, \mathrm{d}y \right) \left(\int_X \|K(x, y)\| \|\phi(y)\|^2 \, \mathrm{d}y \right).$$

By integrating, this implies that

$$\|T\phi\|_{L^2(X, F)}^2 \leq \int_X \left(\int_X \|K(x, y)\| \, \mathrm{d}y \right) \left(\int_X \|K(x, y)\| \|\phi(y)\|^2 \, \mathrm{d}y \right) \mathrm{d}x.$$

This yields $\|T\phi\|_{L^2(X, F)}^2 \leq C_1 C_2 \|\phi\|_{L^2(X, F)}^2$. $\qquad \square$

Composition with Differential Operators

Assume that F is endowed with a Hermitian connection ∇. Let $K \in \mathcal{C}^1(X \times \overline{X}, F \boxtimes \overline{F})$ and let T be the operator with Schwartz kernel K. The two following results allow to compute the kernel of the composition of T with some differential operator.

Lemma 6.4.3. *Let* $Z \in \mathcal{C}^0(X, TX)$ *be a continuous vector field on* X. *Then the Schwartz kernel of* $\nabla_Z \circ T$ *is equal to* $(\nabla_Z \boxtimes \mathrm{id})K$.

Proof. Recall that for $\phi \in \mathcal{C}^0(X, F)$ and $x \in X$,

$$(T\phi)(x) = \int_X K(x, y) \cdot \phi(y) \, \mathrm{d}y.$$

Since X is compact, we can differentiate under the integral sign, which yields

$$((\nabla_Z \circ T)\phi)(x) = \int_X ((\nabla_Z \boxtimes \mathrm{id})K)(x, y) \cdot \phi(y) \, \mathrm{d}y,$$

which was to be proved. $\qquad\qquad\qquad\qquad\qquad\qquad\qquad\qquad\qquad\qquad\quad\square$

Lemma 6.4.4. *Let* $Z \in \mathcal{C}^1(X, TX)$ *be a* $\mathcal{C}^1$ *vector field on* X. *Then the Schwartz kernel of* $T \circ \nabla_Z$ *is equal to* $-(\mathrm{id} \boxtimes (\nabla_Z + \mathrm{div}\, Z))K$. *In this statement, we still use the notation* ∇ *for the induced connection on* $\overline{F}$.

Proof. Set $R := T \circ \nabla_Z$. Let $\phi \in \mathcal{C}^1(X, F)$ and $x \in X$. Then

$$(R\phi)(x) = \int_X K(x, y) \cdot (\nabla_Z \phi)(y) \, \mathrm{d}y.$$

Let us consider the function $f_x = K(x, \cdot) \cdot \phi$. Since ∇ is Hermitian, we have that

$$(\mathcal{L}_Z f_x)(y) = K(x, y) \cdot (\nabla_Z \phi)(y) + ((\mathrm{id} \boxtimes \nabla_Z)K)(x, y) \cdot \phi(y)$$

for every $y \in X$. Integrating this equality, we obtain:

$$(R\phi)(x) = \int_X (\mathcal{L}_Z f_x)(y) \, \mathrm{d}y - \int_X ((\mathrm{id} \boxtimes \nabla_Z)K)(x, y) \cdot \phi(y) \, \mathrm{d}y.$$

Let us change a little bit our notation and call μ the volume form on X, so that

$$(R\phi)(x) = \int_X (\mathcal{L}_Z f_x) \, \mu - \int_X ((\mathrm{id} \boxtimes \nabla_Z)K)(x, y) \cdot \phi(y) \, \mathrm{d}y.$$

Using the Leibniz rule, we have that

$$(\mathcal{L}_Z f_x) \, \mu = \mathcal{L}_Z(f_x \mu) - f_x \mathcal{L}_Z \mu = \mathcal{L}_Z(f_x \mu) - (\mathrm{div}\, Z) f_x \mu$$

where the second equality comes from the definition of the divergence. Since Cartan's formula yields the exactness of $\mathcal{L}_Z(f_x \mu)$, we obtain after integration:

$$\int_X (\mathcal{L}_Z f_x)\,\mu = -\int_X (\operatorname{div} Z) f_x\,\mu$$

and consequently

$$(R\phi)(x) = -\int_X (\operatorname{div} Z) K(x,y) \cdot \phi(y)\,\mathrm{d}y - \int_X \big((\mathrm{id} \boxtimes \nabla_Z) K\big)(x,y) \cdot \phi(y)\,\mathrm{d}y,$$

which yields the result. $\qquad\square$

Chapter 7
Asymptotics of the Projector Π_k

The goal of this chapter is to describe the asymptotic properties of the Schwartz kernel of the Szegő projector $\Pi_k \colon L^2(M, L^k) \to \mathcal{H}_k$, called the Bergman kernel.

7.1 The Section E

Let $\overline{M}$ be the manifold M endowed with the symplectic form $-\omega$ and the complex structure opposite to the complex structure on M. Let $\Delta_M := \operatorname{diag}(M \times \overline{M})$ be the diagonal of $M \times \overline{M}$; observe that it is a Lagrangian submanifold. We want to understand the Schwartz kernel of a Berezin–Toeplitz operator, and it turns out that this kernel concentrates on Δ_M in a certain sense that we will explain later. In order to do so, we introduce some special section of $L \boxtimes \bar{L} \to M \times \overline{M}$ related with Δ_M. We start by introducing some notation, close to the one used in [14], which has been our main inspiration for this section.

Let V be a smooth manifold. We say that a function $f \in \mathcal{C}^\infty(X)$ vanishes to order $N \geq 1$ along a submanifold $Y \subset V$ if for every $m \in [\![0, N-1]\!]$, and for any vector fields $X_1, \ldots, X_m$,

$$\left(\mathcal{L}_{X_1} \cdots \mathcal{L}_{X_m} f \right)_{|Y} = 0.$$

If $K \to V$ is a complex line bundle with connection ∇, we say that a section $s \in \mathcal{C}^\infty(V, K)$ vanishes to order $N \geq 1$ along Y if for every $m \in [\![0, N-1]\!]$, and for any vector fields $X_1, \ldots, X_m$,

$$(\nabla_{X_1} \cdots \nabla_{X_m} s)_{|Y} = 0.$$

A function or section is said to vanish to infinite order along Y if it vanishes to order N along Y for every $N \geq 1$. We will denote by $\mathcal{I}_\infty(Y)$ the set of functions, or sections (the context will solve this ambiguity) vanishing to infinite order along Y.

Proposition 7.1.1. *There exists a section E of $L \boxtimes \bar{L} \to M \times \overline{M}$ such that:*

(1) $\forall x \in M,\ E(x, x) = 1$,

© Springer International Publishing AG, part of Springer Nature 2018
Y. Le Floch, *A Brief Introduction to Berezin–Toeplitz Operators on Compact Kähler Manifolds*, CRM Short Courses,
https://doi.org/10.1007/978-3-319-94682-5_7

(2) *for every $x, y \in M$ with $x \neq y$, $\|E(x,y)\| < 1$,*
(3) *for every $x, y \in M$, $\overline{E(x,y)} = E(y,x)$,*
(4) *for any $Z \in \mathcal{C}^\infty(M, T^{1,0}M)$, the sections $\widetilde{\nabla}_{(\overline{Z},0)}E$ and $\widetilde{\nabla}_{(0,Z)}E$ belong to $\mathcal{I}_\infty(\Delta_M)$.*

Such a section is unique up to an element of $\mathcal{I}_\infty(\Delta_M)$. In this statement, $\widetilde{\nabla}$ is the connection induced by ∇ on $L \boxtimes \overline{L}$, and $\|\cdot\|$ is the norm induced by h on $L \boxtimes \overline{L}$.

The first property is to be understood as follows: $E(x,x)$ is an element of $L_x \otimes \overline{L}_x$, and we identify this bundle with the trivial bundle by means of the metric h. We begin by proving a slightly weaker version of this proposition.

Proposition 7.1.2. *There exists a section E of $L \boxtimes \overline{L} \to M \times \overline{M}$ satisfying points (1) and (4) of the previous proposition. Such a section is unique up to an element of $\mathcal{I}_\infty(\Delta_M)$.*

Proof. Let s be a local non-vanishing holomorphic section on some open subset $U \subset M$, and let $\phi = -\log\big(h(s,s)\big)$, so that $h(s,s) = \exp(-\phi)$. Then $u = \exp(\phi/2)s$ is such that $u(x) \otimes \bar{u}(x) = h_x\big(u(x), u(x)\big) = 1$ for every $x \in U$. So we look for some local section E of $L \boxtimes \overline{L}$ over $U \times U$ of the form

$$E(x,y) = \exp\big(\mathrm{i}\psi(x,y)\big)\, u(x) \otimes \bar{u}(y)$$

with $\psi(x,x) = 0$. Since $u(x) \otimes \bar{u}(y) = \exp\big(\tfrac{1}{2}(\phi(x) + \phi(y))\big) s(x) \otimes \bar{s}(y)$, we have that

$$\widetilde{\nabla}\big(u(x) \otimes \bar{u}(y)\big) = \mathrm{d}\varphi \otimes \big(u(x) \otimes \bar{u}(y)\big) + \exp(\varphi)\widetilde{\nabla}\big(s(x) \otimes \bar{s}(y)\big)$$

where $\varphi(x,y) = \big(\phi(x) + \phi(y)\big)/2$. Therefore,

$$\widetilde{\nabla}E = \mathrm{i}\,\mathrm{d}\widetilde{\psi} \otimes E + \exp(\mathrm{i}\psi + \varphi)\,\widetilde{\nabla}\big(s(x) \otimes \bar{s}(y)\big)$$

with $\widetilde{\psi}(x,y) = \psi(x,y) - (\mathrm{i}/2)\big(\phi(x) + \phi(y)\big)$. Let us introduce some local holomorphic coordinates $z_1, \ldots, z_n$ on U, and let $(z_1^\ell, \ldots, z_n^\ell, z_1^r, \ldots, z_n^r)$ be the corresponding coordinates on $U \times U$. Since s is holomorphic, $\nabla_Z \bar{s} = 0$ for every $Z \in \mathcal{C}^\infty(M, T^{1,0}M)$, and the above computation shows that condition (4) in the statement of the proposition and the fact that $\psi(x,x) = 0$ are equivalent to the equations

$$\widetilde{\psi}(x,x) = -\mathrm{i}\phi(x), \quad \frac{\partial \widetilde{\psi}}{\partial \bar{z}_j^\ell} = 0, \quad \frac{\partial \widetilde{\psi}}{\partial z_j^r} = 0 \bmod \mathcal{I}_\infty\big(\mathrm{diag}(U \times U)\big), \ 1 \leq j \leq n.$$

We claim that there exists a function $\widetilde{\psi}$, unique modulo $\mathcal{I}_\infty\big(\mathrm{diag}(U \times U)\big)$, satisfying these equations. Indeed, these equations force the Taylor expansion of $\widetilde{\psi}$ along the diagonal to be of the form

$$\widetilde{\psi}(w + z^\ell, w + z^r) = -\mathrm{i} \sum_{\alpha,\beta \in \mathbb{N}^n} \frac{1}{\alpha!\beta!}\left(\frac{\partial^{\alpha+\beta}\phi}{\partial z_\alpha^\ell \partial \bar{z}_\beta^r}\right)(w)(z^\ell)^\alpha(\bar{z}^r)^\beta$$

where we have used standard notation from multivariable calculus. This deals with the uniqueness part. For the existence part, the Whitney extension theorem (see e.g. [26, Theorem 2.3.6]) implies that we can construct a function with this given Taylor expansion along the diagonal. Introducing a partition of unity subordinate to a finite cover of Δ_M by such open subsets $U \times U$ and using local uniqueness, we can construct E globally and prove that it is unique modulo an element of $\mathcal{I}_\infty(\Delta_M)$. $\quad\square$

Before showing that E can be chosen to fulfil all the properties required by Proposition 7.1.1, we state further properties of sections given by Proposition 7.1.2.

Lemma 7.1.3. *Let E be as in Proposition 7.1.2, and let us introduce the one-form α_E defined on a neighbourhood of Δ_M by $\widetilde{\nabla} E = -\mathrm{i}\alpha_E \otimes E$. Then*

- *α_E vanishes along Δ_M,*
- *there exists a section B_E of $\left(T^*(M \times \overline{M}) \otimes T^*(M \times \overline{M})\right) \otimes \mathbb{C} \to \Delta_M$ such that for any vector fields X, Y of $M \times \overline{M}$, $\mathcal{L}_X\left(\alpha_E(Y)\right) = B_E(X, Y)$ along Δ_M; moreover, twice the anti-symmetric part of B_E is equal to $\widetilde{\omega} = p_1^*\omega - p_2^*\omega$,*
- *for every $x \in \Delta_M$, for any $X, Y \in T_x(M \times \overline{M}) \otimes \mathbb{C}$,*

$$B_E(X, Y) = \widetilde{\omega}(q(X), Y), \tag{7.1}$$

where q is the projection from $T_x(M \times \overline{M}) \otimes \mathbb{C}$ onto $T_x^{0,1}(M \times \overline{M})$ with kernel $T_x\Delta_M \otimes \mathbb{C}$.

In order to set notation for the proof, let $\tilde{\jmath}$ denote the complex structure on $M \times \overline{M}$ and let $\tilde{g} = \widetilde{\omega}(\cdot, \tilde{\jmath}\cdot)$ be the Kähler metric on $M \times \overline{M}$.

Proof. We have that for any $x \in \Delta_M$, $T_x^{0,1}(M \times \overline{M}) \oplus (T_x\Delta_M \otimes \mathbb{C}) = T_x(M \times \overline{M}) \otimes \mathbb{C}$. Indeed, let $Z \in T_x^{0,1}(M \times \overline{M}) \cap T_x\Delta_M \otimes \mathbb{C}$. Since $T_x\Delta_M \otimes \mathbb{C}$ is Lagrangian and $\overline{Z}$ also belongs to $T_x\Delta_M \otimes \mathbb{C}$, we have that $\widetilde{\omega}(Z, \overline{Z}) = 0$. But since Z belongs to $T_x^{0,1}(M \times \overline{M})$, we can write $Z = X + \mathrm{i}\tilde{\jmath}X$ for some $X \in T_xM$. Therefore

$$0 = \widetilde{\omega}(X + \mathrm{i}\tilde{\jmath}X, X - \mathrm{i}\tilde{\jmath}X) = -2\mathrm{i}\widetilde{\omega}(X, \tilde{\jmath}X).$$

Consequently, $\tilde{g}(X, X) = \widetilde{\omega}(X, \tilde{\jmath}X) = 0$, thus $X = 0$ and $Z = 0$.

Now, for $X \in T_x\Delta_M \otimes \mathbb{C}$, $\alpha_E(X) = 0$ because $E_{|\Delta_M} = 1$, and for $Y \in T_x^{0,1}(M \times \overline{M})$, $\alpha_E(Y) = 0$ because $\widetilde{\nabla}_Y E = 0$. Consequently, $\alpha_E = 0$ along Δ_M.

B_E is well-defined because the value of $\mathcal{L}_X\left(\alpha_E(Y)\right)$ at $x \in \Delta_M$ only depends on the values of X and Y at x. Given two vector fields X, Y of $M \times \overline{M}$, we have

$$\widetilde{\omega}(X, Y) = \mathrm{d}\alpha_E(X, Y) = \mathcal{L}_X\left(\alpha_E(Y)\right) - \mathcal{L}_Y\left(\alpha_E(X)\right) - \alpha_E([X, Y])$$

because the curvature of $\widetilde{\nabla}$ is $-\mathrm{i}\widetilde{\omega}$. Evaluating at a point on Δ_M and remembering that α_E vanishes along Δ_M, we obtain

$$\widetilde{\omega}(X, Y) = B_E(X, Y) - B_E(Y, X).$$

Since α_E vanishes along Δ_M, we have that for $x \in \Delta_M$, $X \in T_x\Delta_M \otimes \mathbb{C}$ and $Y \in T_x(M \times \overline{M}) \otimes \mathbb{C}$, $B_E(X,Y) = 0$; hence

$$\forall x \in \Delta_M, \forall X, Y \in T_x(M \times \overline{M}) \otimes \mathbb{C}, \quad B_E(X,Y) = B_E(q(X), Y).$$

Therefore, it suffices to prove that $B_E(X,Y) = \widetilde{\omega}(X,Y)$ whenever X belongs to $T_x^{0,1}(M \times \overline{M})$. But we know that for any $X \in T^{0,1}(M \times \overline{M})$ and for any $Y \in T(M \times \overline{M}) \otimes \mathbb{C}$, $\widetilde{\nabla}_Y \widetilde{\nabla}_X E = 0$ on Δ_M, which yields

$$\forall x \in \Delta_M, \forall X \in T_x^{0,1}(M \times \overline{M}), \forall Y \in T_x(M \times \overline{M}) \otimes \mathbb{C}, \quad B_E(Y,X) = 0.$$

We conclude by using that twice the anti-symmetric part of B_E is $\widetilde{\omega}$. $\qquad\square$

Lemma 7.1.4. *We consider E as in Proposition 7.1.2 and introduce the function $\varphi_E := -2\log\|E\|$. Then $\mathrm{d}\varphi_E$ vanishes along Δ_M and the Hessian of φ_E at $x \in \Delta_M$ is the bilinear symmetric form of $T_x(M \times \overline{M})$ with kernel $T_x\Delta_M$ and whose restriction to $\widetilde{\jmath}(T_x\Delta_M)$ is equal to $2\widetilde{g}$.*

Proof. Using the compatibility between $\widetilde{\nabla}$ and the Hermitian metric on $L \boxtimes \bar{L}$, we find that $\mathrm{d}\|E\|^2 = -\mathrm{i}(\alpha_E - \overline{\alpha_E})\|E\|^2$, so

$$\mathrm{d}\varphi_E = \mathrm{i}(\alpha_E - \overline{\alpha_E}).$$

Hence $\mathrm{d}\varphi_E$ vanishes along Δ_M because α_E itself does. Let us compute the Hessian of φ_E at $X, Y \in T_x(M \times \overline{M})$, which is defined as

$$\mathrm{Hess}_{\varphi_E}(x)(X,Y) = \mathcal{L}_{\widehat{X}}(\mathcal{L}_{\widehat{Y}}\varphi_E)(x)$$

for any vector fields $\widehat{X}, \widehat{Y}$ such that $\widehat{X}(x) = X$ and $\widehat{Y}(x) = Y$. By the computation above, we obtain that

$$\mathrm{Hess}_{\varphi_E}(x)(X,Y) = \mathrm{i}\mathcal{L}_{\widehat{X}}\big(\alpha_E(\widehat{Y}) - \overline{\alpha_E}(\widehat{Y})\big)(x) = \mathrm{i}\big(B_E(X,Y) - \overline{B_E(X,Y)}\big).$$

Since $B_E(X,Y) = 0$ whenever X belongs to $T_x\Delta_M$, the kernel of this Hessian contains $T_x\Delta_M$. Furthermore, since $T_x\Delta_M$ is Lagrangian, its orthogonal complement with respect to $\widetilde{g}$ is $\widetilde{\jmath}(T_x\Delta_M)$. But if X belongs to $\widetilde{\jmath}(T_x\Delta_M)$, we have $q(X) = X + \mathrm{i}\widetilde{\jmath}X$ because $X = X + \mathrm{i}\widetilde{\jmath}X - \mathrm{i}\widetilde{\jmath}X$, $X + \mathrm{i}\widetilde{\jmath}X$ belongs to $T_x^{0,1}(M \times \overline{M})$ and $-\mathrm{i}\widetilde{\jmath}X$ belongs to $T_x\Delta_M \otimes \mathbb{C}$. Thus using formula (7.1) we get that

$$\mathrm{Hess}_{\varphi_E}(X,Y) = -2\widetilde{\omega}(\widetilde{\jmath}X,Y) = 2\widetilde{g}(X,Y).$$

for $X, Y \in \widetilde{\jmath}(T_x\Delta_M)$. $\qquad\square$

We are now ready to complete the proof of Proposition 7.1.1.

Proof of Proposition 7.1.1. Pick any E as in Proposition 7.1.2. The previous lemma shows that the Hessian of $\varphi_E = -2\log\|E\|$ is positive on the orthogonal complement of $T\Delta_M$. Hence, there exists a neighbourhood U of Δ_M such that φ_E itself is positive on $U \setminus \Delta_M$. Thus, modifying E outside U if necessary, we obtain that $\|E\| < 1$ outside Δ_M. Now, observe that if E satisfies conditions (1), (2), (4) of Proposition 7.1.1, the section $(x, y) \mapsto \frac{1}{2}(E(x, y) + \overline{E(y, x)})$ satisfies conditions (1), (2), (3), (4) of this proposition. $\qquad\square$

Example 7.1.5 (*On the plane*). Let $M = \mathbb{R}^2$ and $L = M \times \mathbb{C} \to M$ as before. Define $E \in \mathcal{C}^\infty(M \times \overline{M}, L \boxtimes \overline{L})$ as

$$E(z, w) = \exp(z\overline{w})\, \psi(z) \otimes \overline{\psi}(w)$$

with $\psi(z) = \exp(-|z|^2/2)$. From this expression, it is easy to check that the properties of E agree with the ones listed in Proposition 7.1.1; indeed, $\overline{E(z, w)} = E(w, z)$, E is holomorphic with respect to z and anti-holomorphic with respect to w, and

$$\|E(z, w)\|^2 = \exp(-|z - w|^2)$$

is equal to one when $z = w$ and strictly smaller than one when $z \neq w$. The function φ_E is given by $\varphi_E(z, w) = |z - w|^2$ and satisfies the properties stated in Lemma 7.1.4. We can also compute the differential form α_E and the bilinear form B_E. Indeed, we have that

$$\widetilde{\nabla} E = (\overline{w}\, \mathrm{d}z + z\, \mathrm{d}\overline{w} - \overline{z}\, \mathrm{d}z - w\, \mathrm{d}\overline{w}) \otimes E$$

since $\nabla \psi = -\overline{z}\, \mathrm{d}z \otimes \psi$. Consequently, we obtain that

$$\alpha_E = \mathrm{i}(\overline{w} - \overline{z})\, \mathrm{d}z + \mathrm{i}(z - w)\, \mathrm{d}\overline{w},$$

and it follows that

$$B_E(X, Y) = \big(\mathrm{d}\overline{w}(X) - \mathrm{d}\overline{z}(X)\big)\, \mathrm{d}z(Y) + \big(\mathrm{d}z(X) - \mathrm{d}w(X)\big)\, \mathrm{d}\overline{w}(Y).$$

In particular, we have that $B_E(X, Y) - B_E(Y, X) = \widetilde{\omega}(X, Y)$ as expected, and

$$B_E(X, Y) + B_E(Y, X) = 2\mathrm{i}\big(\mathrm{d}z \otimes (\mathrm{d}\overline{w} - \mathrm{d}\overline{z}) + \mathrm{d}\overline{w} \otimes (\mathrm{d}z - \mathrm{d}w)\big)(X, Y).$$

Example 7.1.6 (*The unit disc*). We consider the unit disc as in Example 4.4.3, and as in this example we set $\psi(z) = \sqrt{1 - |z|^2}$. We claim that

$$E(z, w) = \frac{1}{1 - z\overline{w}} \psi(z) \otimes \overline{\psi}(w)$$

satisfies the required properties. Indeed, E is holomorphic in z and anti-holomorphic in w, and $E(w, z) = \overline{E(z, w)}$. Moreover, we have that

$$\|E(z, w)\|^2 = \frac{(1 - |z|^2)(1 - |w|^2)}{|1 - z\overline{w}|^2},$$

which is equal to one when $z = w$ and strictly smaller than one otherwise. Indeed, when $z \neq w$, we have that $2\, \Re(z\overline{w}) < |z|^2 + |w|^2$ because $|z - w|^2 > 0$.

Exercise 7.1.7. Compute φ_E, α_E and β_E for this example, and check that they satisfy the conclusions of Lemma 7.1.3.

Example 7.1.8 (*Complex projective spaces*). Remember that on $M = \mathbb{CP}^n$ endowed with the Fubini–Study symplectic form, the dual bundle $\mathcal{O}(1)$ of the tautological bundle is a prequantum line bundle. Let $U_0, \ldots, U_n$ be the trivialisation open sets and $s_0, \ldots, s_n$ be the associated unit sections of $\mathcal{O}(-1)$, as defined in Example 4.4.5. For $j \in [\![0, n]\!]$, let t_j be the local section of L which is dual to s_j. Introduce also the usual holomorphic coordinates on each U_j, and define

$$E(z, w) = (1 + \langle z, w \rangle)\, t_j(z) \otimes \bar{t}_j(w),$$

where $\langle \cdot, \cdot \rangle$ stands for the usual Hermitian product on $\mathbb{C}^n$; let $\|\cdot\|_{\mathbb{C}^n}$ be the associated norm. We claim that E satisfies the properties stated in Proposition 7.1.1. It is clear that $E(z, z) = 1$, and the norm of E satisfies

$$\|E(z, w)\|^2 = \frac{|1 + \langle z, w \rangle|^2}{(1 + \|w\|^2_{\mathbb{C}^n})(1 + \|z\|^2_{\mathbb{C}^n})}$$

since $h(s_j, s_j)(z) = 1 + \|z\|^2_{\mathbb{C}^n}$. This quantity is strictly smaller than one whenever $z \neq w$ because of the Cauchy–Schwarz inequality. Finally, observe that t_j is holomorphic and the function $(z, w) \mapsto 1 + \langle z, w \rangle$ is holomorphic with respect to z and anti-holomorphic with respect to w. We can also compute the differential form α_E. Since $\nabla t_j = -\partial \phi_j \otimes t_j$ where ϕ_j is the Kähler potential on U_j introduced in Example 2.5.9, we get that

$$\alpha_E = \mathrm{i}\left(\frac{\mathrm{d}f}{f} - \partial \phi_j(z) - \bar{\partial} \phi_j(w) \right)$$

where $f(z, w) = 1 + \langle z, w \rangle$. This reads

$$\alpha_E = \mathrm{i}\left(\frac{\sum_{m=1}^{n}(z_m\, \mathrm{d}\overline{w}_m + \overline{w}_m\, \mathrm{d}z_m)}{1 + \sum_{m=1}^{n} z_m \overline{w}_m} - \frac{\sum_{m=1}^{n} \bar{z}_m\, \mathrm{d}z_m}{1 + \sum_{m=1}^{n}|z_m|^2} - \frac{\sum_{m=1}^{n} w_m\, \mathrm{d}\overline{w}_m}{1 + \sum_{m=1}^{n}|w_m|^2} \right),$$

hence α_E indeed vanishes on the diagonal of $M \times M$. Since $\alpha_E(\partial_{\bar{z}_\ell})$ and $\alpha_E(\partial_{w_\ell})$ vanish, we have that

$$B_E(X, \partial_{\bar{z}_\ell}) = 0 = B_E(X, \partial_{w_\ell})$$

for every X. By differentiating the expression

$$\alpha_E(\partial_{z_\ell}) = \mathrm{i}\left(\frac{\overline{w}_\ell}{1 + \sum_{m=1}^{n} z_m \overline{w}_m} - \frac{\bar{z}_\ell}{1 + \sum_{m=1}^{n}|z_m|^2} \right)$$

and evaluating at the point (z, z) of the diagonal, we obtain that

$$B_E(\partial_{z_p}, \partial_{z_\ell}) = 0 = B_E(\partial_{w_p}, \partial_{z_\ell}), \quad B_E(\partial_{\bar{z}_p}, \partial_{z_\ell}) = -\mathrm{i}\left(\frac{(1 + \|z\|^2_{\mathbb{C}^n})\delta_{\ell, p} - z_p \bar{z}_\ell}{(1 + \|z\|^2_{\mathbb{C}^n})^2} \right)$$

and also

$$B_E(\partial_{\overline{w}_p}, \partial_{z_\ell}) = \mathrm{i}\left(\frac{(1+\|z\|_{\mathbb{C}^n}^2)\delta_{\ell,p} - z_p\overline{z}_\ell}{(1+\|z\|_{\mathbb{C}^n}^2)^2}\right).$$

Similar computations yield

$$B_E(\partial_{\overline{w}_p}, \partial_{\overline{w}_\ell}) = 0 = B_E(\partial_{\overline{z}_p}, \partial_{\overline{w}_\ell}), \quad B_E(\partial_{w_p}, \partial_{\overline{w}_\ell}) = -\mathrm{i}\left(\frac{(1+\|z\|_{\mathbb{C}^n}^2)\delta_{\ell,p} - z_\ell\overline{z}_p}{(1+\|z\|_{\mathbb{C}^n}^2)^2}\right)$$

and finally

$$B_E(\partial_{z_p}, \partial_{\overline{w}_\ell}) = \mathrm{i}\left(\frac{(1+\|z\|_{\mathbb{C}^n}^2)\delta_{\ell,p} - z_\ell\overline{z}_p}{(1+\|z\|_{\mathbb{C}^n}^2)^2}\right).$$

Hence, we finally obtain that

$$B_E(X,Y)$$

$$= \frac{\mathrm{i}}{(1+\|z\|_{\mathbb{C}^n}^2)^2} \sum_{\ell,p=1}^n \left((1+\|z\|_{\mathbb{C}^n}^2)\delta_{\ell,p} - z_p\overline{z}_\ell\right)\left(\mathrm{d}\overline{w}_p(X)\,\mathrm{d}z_\ell(Y) - \mathrm{d}\overline{z}_p(X)\,\mathrm{d}z_\ell(Y)\right)$$

$$+ \frac{\mathrm{i}}{(1+\|z\|_{\mathbb{C}^n}^2)^2} \sum_{\ell,p=1}^n \left((1+\|z\|_{\mathbb{C}^n}^2)\delta_{\ell,p} - z_\ell\overline{z}_p\right)\left(\mathrm{d}z_p(X)\,\mathrm{d}\overline{w}_\ell(Y) - \mathrm{d}w_p(X)\,\mathrm{d}\overline{w}_\ell(Y)\right),$$

so the map $(X,Y) \mapsto B_E(X,Y) - B_E(Y,X)$ coincides with

$$\frac{\mathrm{i}}{(1+\|z\|_{\mathbb{C}^n}^2)^2} \sum_{\ell,p=1}^n \left((1+\|z\|_{\mathbb{C}^n}^2)\delta_{\ell,p} - z_p\overline{z}_\ell\right)(\mathrm{d}z_\ell \wedge \mathrm{d}\overline{z}_p - \mathrm{d}w_\ell \wedge \mathrm{d}\overline{w}_m) = \widetilde{\omega},$$

as expected.

Example 7.1.9 (*Two-dimensional symplectic vector space*). We come back to Example 4.4.8 and keep the same notation. Namely, $\tau = a + \mathrm{i}b$ parametrises the complex structure on V, and we work with the complex coordinate $z = p + \tau q$. We consider the section

$$E(z,w) = \exp\left(-\frac{\pi}{b}(z - \overline{w})^2\right)t(z) \otimes \overline{t}(w),$$

where t was the holomorphic section defined in (4.2). The facts that E is holomorphic in z and anti-holomorphic in w and that $E(w,z) = \overline{E(z,w)}$ are obvious. Moreover, since $\|t(p,q)\|^2 = \exp(-4\pi bq^2)$, we have that $E(z,z) = 1$. Finally, a straightforward computation shows that

$$\|E(z,w)\|^2 = \exp\left(-\frac{2\pi}{b}|z - w|^2\right),$$

and this quantity is strictly smaller than one when $z \neq w$.

7.2 Schwartz Kernel of the Projector

The following theorem is the most fundamental result in the theory of Berezin–Toeplitz operators. It describes the Schwartz kernel of the Szegő projector and is essential to derive most of the crucial properties of these operators.

Theorem 7.2.1 ([14, 33]). *The projector Π_k has a Schwartz kernel of the form*

$$\Pi_k(x, y) = \left(\frac{k}{2\pi}\right)^n E^k(x, y) u(x, y, k) + R_k(x, y)$$

where the section E is as in Proposition 7.1.1, $u(\cdot, \cdot, k)$ is a sequence of functions in $C^\infty(M \times M, \mathbb{R})$ having an asymptotic expansion of the form

$$u(\cdot, \cdot, k) \sim \sum_{\ell \geq 0} k^{-\ell} u_\ell(\,\cdot\,, \cdot\,)$$

for the C^∞-topology, with $u_0(x, x) = 1$, where for any $Z \in C^\infty(M, T^{1,0}M)$, the functions $\mathcal{L}_{(\overline{Z},0)} u_\ell$ and $\mathcal{L}_{(0,Z)} u_\ell$ vanish to infinite order along the diagonal Δ_M of M^2, and $R_k = O(k^{-\infty})$ uniformly in (x, y).

Here, the meaning of $u(\cdot, \cdot, k) \sim \sum_{\ell \geq 0} k^{-\ell} u_\ell(\,\cdot\,, \cdot\,)$ is that for every $N \geq 0$, the function $u(\cdot, \cdot, k) - \sum_{\ell=0}^{N} k^{-\ell} u_\ell(\cdot, \cdot)$ and all its derivatives are uniformly $O(k^{-(N+1)})$. From this result, we can recover the equivalent of the dimension of $\mathcal{H}_k$ stated in Theorem 4.2.4. Indeed, Proposition 6.3.4 yields

$$\dim \mathcal{H}_k = \mathrm{Tr}(\Pi_k) = \int_M \Pi_k(x, x) \, \mathrm{d}\mu(x) = \left(\frac{k}{2\pi}\right)^n \mathrm{vol}(M) + O(k^{n-1}).$$

The rest of this section will be devoted to sketching a proof of Theorem 7.2.1. Before doing so, let us give some examples.

Example 7.2.2 (The plane). Remember that the Hilbert space at level k in the quantisation of the plane is

$$\mathcal{H}_k = \left\{ f\psi^k \,\middle|\, f \colon \mathbb{C} \to \mathbb{C} \text{ holomorphic}, \int_{\mathbb{C}} |f(z)|^2 \exp(-k|z|^2) \,|\, \mathrm{d}z \wedge \mathrm{d}\bar{z}| < +\infty \right\}$$

with Hermitian inner product

$$\left\langle f\psi^k, g\psi^k \right\rangle_k = \int_{\mathbb{C}} f(z)\bar{g}(z) \exp(-k|z|^2) \,|\, \mathrm{d}z \wedge \mathrm{d}\bar{z}|.$$

One can deduce from the fact that holomorphic functions are analytic that the monomials $(z^n \psi^k)_{n \geq 0}$ generate $\mathcal{H}_k$. Furthermore, using polar coordinates (ρ, θ), we have that

$$\langle z^m \psi^k, z^n \psi^k \rangle_k = 2 \left(\int_0^{2\pi} \exp(\mathrm{i}(m-n)\theta) \, \mathrm{d}\theta \right) \left(\int_0^{+\infty} \rho^{m+n+1} \exp(-k\rho^2) \, \mathrm{d}\rho \right)$$

which is zero whenever $m \neq n$, which means that the monomials form an orthogonal basis of $\mathcal{H}_k$. Their norm satisfies

$$\| z^n \psi^k \|_k^2 = 4\pi \left(\int_0^{+\infty} \rho^{2n+1} \exp(-k\rho^2) \, \mathrm{d}\rho \right).$$

A straightforward computation using integration by parts yields that the integral on the right-hand side is equal to $n!/(2k^{n+1})$, hence the family

$$\phi_{k,n} = \sqrt{\frac{k^{n+1}}{2\pi n!}} \, z^n \psi^k, \quad n \geq 0$$

is an orthonormal basis of $\mathcal{H}_k$. Consequently, the action of the projector Π_k on $\varphi \in L^2(M, L^k)$ is given by the formula

$$\Pi_k \varphi = \sum_{n \geq 0} \langle \varphi, \phi_{k,n} \rangle_k \phi_{k,n}.$$

Therefore

$$(\Pi_k \varphi)(z) = \sum_{n \geq 0} \frac{k^{n+1}}{2\pi n!} \left(\int_{\mathbb{C}} \varphi(w) \overline{w}^n \exp\left(-\frac{k}{2} |w|^2 \right) \mathrm{i} \, \mathrm{d}w \wedge \mathrm{d}\overline{w} \right) z^n \exp\left(-\frac{k}{2} |z|^2 \right).$$

Interchanging the sum and the integral (exercise: justify this!), this yields

$$(\Pi_k \varphi)(z) = \int_{\mathbb{C}} \frac{k}{2\pi} \left(\sum_{n \geq 0} \frac{(kz\overline{w})^n}{n!} \right) \exp\left(-\frac{k}{2} (|w|^2 + |z|^2) \right) \varphi(w) \, \mathrm{i} \, \mathrm{d}w \wedge \mathrm{d}\overline{w},$$

and finally

$$(\Pi_k \varphi)(z) = \int_{\mathbb{C}} \frac{k}{2\pi} \exp(z\overline{w}) \exp\left(-\frac{k}{2} (|w|^2 + |z|^2) \right) \varphi(w) \, \mathrm{i} \, \mathrm{d}w \wedge \mathrm{d}\overline{w}.$$

This means that the kernel of the projector is equal to

$$\Pi_k(z, w) = \frac{k}{2\pi} \exp\left(-\frac{k}{2} (|z|^2 + |w|^2 - 2z\overline{w}) \right) = \frac{k}{2\pi} E^k(z, w),$$

where E is as in Example 7.1.5. Here u is identically 1 and R_k vanishes.

Example 7.2.3 (The unit disc). On the unit disc, keeping the notation of Example 4.4.3, it is easily seen that the family $(z^\ell \psi^k)_{\ell \in \mathbb{N}}$ is an orthogonal basis of $\mathcal{H}_k$, and that the square of the norm of $z^\ell \psi^k$ is equal to

$$I_{k,\ell} = 4\pi \int_0^1 \rho^{2\ell+1}(1-\rho^2)^{k-2}\,\mathrm{d}\rho,$$

by using polar coordinates.

Exercise 7.2.4. Prove that

$$I_{k,\ell} = \frac{2\pi}{(k-1)\binom{k+\ell-1}{\ell}}$$

for any two integers $k \geq 1$ and $\ell \geq 0$.

By using the orthonormal basis that we obtain in this way, a straightforward computation shows that

$$\Pi_k(z,w) = \left(\frac{k-1}{2\pi}\right)\left(\sum_{\ell=0}^{+\infty}\binom{k+\ell-1}{\ell}(z\overline{w})^\ell\right)\psi^k(z)\otimes\overline{\psi}^k(w) = \left(\frac{k-1}{2\pi}\right)E^k(z,w),$$

where E is as in Example 7.1.6. The last equality comes from the relation

$$\frac{1}{(1-u)^k} = \sum_{\ell=0}^{+\infty}\binom{k+\ell-1}{\ell}u^\ell,$$

sometimes called negative binomial theorem, which is valid whenever $|u| < 1$.

Example 7.2.5 (*The complex projective line*). We recall that on $\mathbb{CP}^1$ equipped with the Fubini–Study symplectic form, the line bundle $L = \mathcal{O}(1)$ is a prequantum line bundle, and the Hilbert space $\mathcal{H}_k$ can be identified with the space $\mathbb{C}_k[z_1, z_2]$ of degree k homogeneous polynomials in two complex variables. We should also explain what the scalar product on $\mathcal{H}_k$ becomes through this isomorphism. Let the open sets U_j and the local sections t_j, $j = 0, 1$, be as in Example 7.1.8. We shall denote by z the local coordinate on either U_0 or U_1. To a polynomial $P \in \mathbb{C}_k[z_1, z_2]$, we associate the local sections $P(1,z)t_0^k(z)$ and $P(z,1)t_1^k(z)$ of L^k. Let us work for instance in U_0; the Fubini–Study form is expressed as

$$\omega_{\mathrm{FS}} = \frac{\mathrm{i}\mathrm{d}z \wedge \mathrm{d}\overline{z}}{(1+|z|^2)^2}$$

on this open set. The scalar product of P and Q is thus given by

$$\langle P, Q \rangle_k = \int_{\mathbb{C}} P(1,z)\overline{Q(1,z)}\,h_k(t_0^k, t_0^k)(z)\frac{|\mathrm{d}z \wedge \mathrm{d}\overline{z}|}{(1+|z|^2)^2}.$$

Indeed, observe that U_0 is $\mathbb{CP}^1$ minus a point. Since $h(t_0, t_0)(z) = \left(1+|z|^2\right)^{-1}$, we finally obtain that

$$\langle P, Q \rangle_k = \int_{\mathbb{C}} \frac{P(1,z)\overline{Q(1,z)}}{(1+|z|^2)^{k+2}}\,|\mathrm{d}z \wedge \mathrm{d}\overline{z}|.$$

A basis of $\mathbb{C}_k[z_1, z_2]$ is given by the monomials $f_\ell = z_1^\ell z_2^{k-\ell}, 0 \leq \ell \leq k$. This basis is in fact orthogonal; indeed, using polar coordinates $z = \rho \exp(i\theta)$, we get that

$$\langle f_\ell, f_m \rangle_k = 2 \left(\int_0^{2\pi} \exp(i(\ell - m)\theta) \, d\theta \right) \left(\int_0^{+\infty} \frac{\rho^{2k-(\ell+m)+1}}{(1 + \rho^2)^{k+2}} \, d\rho \right),$$

which vanishes when $\ell \neq m$.

Exercise 7.2.6. Prove that for $0 \leq \ell \leq k$,

$$\int_0^{+\infty} \frac{\rho^{2(k-\ell)+1}}{(1 + \rho^2)^{k+2}} \, d\rho = \frac{1}{2(k+1)\binom{k}{\ell}}.$$

Hint: Look for a primitive of the integrand of the form $P(\rho^2)/(1 + \rho^2)^{k+1}$ with P polynomial of degree $k - \ell$, or use your knowledge of special functions.

The exercise shows that the polynomials

$$e_\ell = \sqrt{\frac{(k+1)\binom{k}{\ell}}{2\pi}} \, z_1^\ell z_2^{k-\ell}, \quad 0 \leq \ell \leq k,$$

form an orthonormal basis of $\mathbb{C}_k[z_1, z_2]$. Therefore, we know from Lemma 6.3.2 that the Schwartz kernel of the projector satisfies

$$\Pi_k(z, w) = \sum_{\ell=0}^k e_\ell(1, z) t_0^k(z) \otimes \overline{e_\ell(1, w)} \, \bar{t}_0^k(w) = \frac{k+1}{2\pi} \sum_{\ell=0}^k \binom{k}{\ell} (z\overline{w})^{k-\ell} t_0^k(z) \otimes \bar{t}_0^k(w),$$

which finally yields

$$\Pi_k(z, w) = \frac{k+1}{2\pi} (1 + z\overline{w})^k \, t_0^k(z) \otimes \bar{t}_0^k(w) = \frac{k}{2\pi} E^k(z, w) \left(1 + \frac{1}{k} \right)$$

where E is as in Example 7.1.8 for $n = 1$. Here $u_0 = 1 = u_1$ and $R_k = 0$. We can check that similarly,

$$\Pi_k(z', w') = \frac{k+1}{2\pi} (1 + z'\overline{w}')^k \, t_1^k(z') \otimes \bar{t}_1^k(w')$$

where z' is the usual holomorphic coordinate on U_1.

Exercise 7.2.7. By using the same reasoning, prove that the kernel of the projector Π_k for $(\mathbb{CP}^n, \omega_{FS})$, endowed with the dual of the tautological line bundle, is given by the formula

$$\Pi_k(z, w) = \frac{(k+n)!}{(2\pi)^n k!} E^k(z, w) = \left(\frac{k}{2\pi} \right)^n E^k(z, w) \prod_{p=1}^n \left(1 + \frac{p}{k} \right)$$

where E is the section defined in Example 7.1.8. *Hint*: Check that the family

$$\frac{1}{(2\pi)^{n/2}}\sqrt{\frac{(k+n)!}{\alpha!}}\, z_0^{\alpha_0} z_1^{\alpha_1} \cdots z_n^{\alpha_n}, \quad \alpha \in \mathbb{N}^{n+1}, \ \alpha_0 + \cdots + \alpha_n = k,$$

forms an orthonormal basis of $\mathcal{H}_k$.

Now that we have an explicit expression for the projector Π_k, let us prove the claims in Example 5.2.4. We will derive the expression for $T_k(x_3)$ and leave the other two as an exercise. Recall that working in the trivialisation U_1 corresponds to sending $P(z_1, z_2)$ to the local section $P(z, 1)t_1^k(z)$. Let $\phi = ft_1^k$, $f \colon \mathbb{C} \to \mathbb{C}$, be a local section in $L^2(M, L^k)$. Then

$$(\Pi_k \phi)(z) = \frac{k+1}{2\pi}\left(\int_{\mathbb{C}} (1+z\overline{w})^k f(w) h_k\big(t_1^k(w), t_1^k(w)\big) \frac{\mathrm{i}\mathrm{d}w \wedge \mathrm{d}\overline{w}}{(1+|w|^2)^2}\right) t_1^k(z),$$

which can be rewritten as

$$(\Pi_k \phi)(z) = \frac{k+1}{2\pi}\left(\int_{\mathbb{C}} \frac{(1+z\overline{w})^k}{(1+|w|^2)^{k+2}} f(w)\, \mathrm{i}\mathrm{d}w \wedge \mathrm{d}\overline{w}\right) t_1^k(z).$$

In particular, since we have that

$$\big((\pi_N^{-1})^* x_3\big)(z) = \frac{|z|^2 - 1}{|z|^2 + 1},$$

then for $\phi = p\,t_1^k$, $p \in \mathbb{C}[z]$ of degree at most k, $T_k(x_3)\phi = q\,t_1^k$ with

$$q(z) = \frac{k+1}{2\pi}\int_{\mathbb{C}} \frac{(1+z\overline{w})^k}{(1+|w|^2)^{k+2}}\left(\frac{|w|^2-1}{|w|^2+1}\right) p(w)\, \mathrm{i}\mathrm{d}w \wedge \mathrm{d}\overline{w}.$$

We want to compare q with the polynomial $z(\mathrm{d}p/\mathrm{d}z)$. Since the latter is a polynomial of degree at most k, we have that

$$z\frac{\mathrm{d}p}{\mathrm{d}z} = \frac{k+1}{2\pi}\int_{\mathbb{C}} \frac{(1+z\overline{w})^k}{(1+|w|^2)^{k+2}}\, w\frac{\mathrm{d}p}{\mathrm{d}w}(w)\, \mathrm{i}\mathrm{d}w \wedge \mathrm{d}\overline{w}.$$

Observe that

$$\frac{\mathrm{d}}{\mathrm{d}w}\left(\frac{w}{(1+|w|^2)^{k+2}}\right) = \frac{1 - (k+1)|w|^2}{(1+|w|^2)^{k+3}}.$$

Exercise 7.2.8. Prove the following formula:

$$\int_{\mathbb{C}} \frac{(1+z\overline{w})^k}{(1+|w|^2)^{k+2}}\, w\frac{\mathrm{d}p}{\mathrm{d}w}(w)\, \mathrm{i}\mathrm{d}w \wedge \mathrm{d}\overline{w}$$
$$= \int_{\mathbb{C}} \frac{(1+z\overline{w})^k}{(1+|w|^2)^{k+2}}\left(\frac{(k+1)|w|^2 - 1}{1+|w|^2}\right) p(w)\, \mathrm{i}\mathrm{d}w \wedge \mathrm{d}\overline{w}.$$

Hint: Apply Stokes' formula in the ball of radius R centred at the origin, and study what happens in the limit $R \to +\infty$.

This implies that $z(\mathrm{d}p/\mathrm{d}z) = T_k\big(g(\cdot, k)\big)p$ with

$$g(z, k) = \frac{(k+1)|z|^2 - 1}{1 + |z|^2} = \frac{|z|^2 - 1}{1 + |z|^2} + k\frac{|z|^2}{1 + |z|^2},$$

which yields in terms of x_3

$$g(\cdot, k) = (\pi_N^{-1})^* x_3 + \frac{k}{2}\big((\pi_N^{-1})^* x_3 + 1\big).$$

Therefore we obtain that

$$z\frac{\mathrm{d}}{\mathrm{d}z} = \frac{k+2}{2}T_k(x_3) + \mathrm{Id},$$

and finally

$$T_k(x_3) = \frac{1}{k+2}\left(2z\frac{\mathrm{d}}{\mathrm{d}z} - k\,\mathrm{Id}\right).$$

Exercise 7.2.9. Prove the formulas for $T_k(x_1)$ and $T_k(x_2)$.

Example 7.2.10 (*Two-dimensional tori*). Let us investigate the case of tori as in Examples 4.4.8 and 5.2.7. We will exploit (4.6) to construct an orthonormal basis of $\mathcal{H}_k^{\mu,\nu}$. We could proceed in several different ways. For instance we could compute the scalar product induced on $\mathbb{R}^{2k}$ by the isomorphism sending an element of $\mathcal{H}_k^{\mu,\nu}$ to its coefficients $(\rho_0, \ldots, \rho_{2k-1})$, and choose a corresponding orthonormal basis on $\mathbb{R}^{2k}$. We will use a different approach, based on the operators $T_{\lambda/2k}^*$ introduced in Example 5.2.7. Firstly, observe that these operators are unitary. Secondly, if (p, q) are coordinates associated with (e, f) as before, then

$$(T_{e/2k}^*\psi)(p, q) = \exp(-\mathrm{i}\pi q)\psi\left(p + \frac{1}{2k}, q\right), \quad (T_{f/2k}^*\psi)(p, q) = \exp(\mathrm{i}\pi p)\psi\left(p, q + \frac{1}{2k}\right)$$

and we deduce from these formulas that

$$T_{e/2k}^* T_{f/2k}^* = \exp\left(\frac{\mathrm{i}\pi}{k}\right)T_{f/2k}^* T_{e/2k}^*.$$

Consequently, if λ_0 is an eigenvalue of $T_{e/2k}^*$ and ψ_0 is a unit eigenvector associated with λ, then $T_{f/2k}^*\psi_0$ is an eigenvector for $T_{e/2k}^*$ with eigenvalue $\exp(\mathrm{i}\pi/k)\lambda_0$. Therefore, $T_{e/2k}^*$ has $2k$ distinct eigenvalues

$$\lambda_0, \lambda_1 = \exp(\mathrm{i}\pi/k)\lambda_0, \ldots, \lambda_{2k-1} = \exp\big((2k-1)\mathrm{i}\pi/k\big)\lambda_0$$

and $T_{f/2k}^*$ sends the eigenspace associated with λ_ℓ to the one associated with $\lambda_{\ell+1}$, for $\ell \in \mathbb{Z}/2k\mathbb{Z}$. Consequently,

$$\psi_0, \psi_1 = T_{f/2k}^*\psi_0, \ldots, \psi_{2k-1} = \big(T_{f/2k}^*\big)^{2k-1}\psi_0$$

forms an orthonormal basis of $\mathcal{H}_k^{\mu,\nu}$. So we only need to find such a pair (λ_0, ψ_0). In order to do so, we consider the function g_0 defined by its coefficients $\rho_0 = 1$ and $\rho_n = 0$ for $1 \le n \le 2k - 1$ as in (4.5), and the associated section $\phi_0 = g_0 t^k$. By (4.7), this gives $\rho_{2mk} = \exp\big(2im\pi(\mu\tau + km\tau - \nu)\big)$ for $m \in \mathbb{Z}$, thus

$$g_0(z) = \exp(2i\pi\mu z) \sum_{m \in \mathbb{Z}} \exp\big(2im\pi(2kz + \mu\tau + km\tau - \nu)\big).$$

One readily checks that

$$g_0\left(z + \frac{1}{2k}\right) = \exp\left(\frac{i\pi\mu}{k}\right) g_0(z).$$

Since moreover $T_{e/2k}^* t^k = t^k$, we obtain that $T_{e/2k}^* \phi_0 = \exp(i\pi\mu/k)\phi_0$. Therefore we get an orthonormal basis by applying the above construction with $\psi_0 = \phi_0/\|\phi_0\|_k$.

Lemma 7.2.11. *As before, let $b = \Im\tau = 4\pi/\omega(e, je) > 0$. Then*

$$\|\phi_0\|_k^2 = \frac{2\pi \exp(\pi b\mu^2/k)}{\sqrt{bk}}.$$

Proof. Recall that $z = p + \tau q$ where (p, q) are coordinates associated with (e, f). We have that

$$\|\phi_0\|_k^2 = 4\pi \int_0^1 \int_0^1 |g_0(p, q)|^2 \exp(-4k\pi bq^2)\, dp\, dq.$$

One can check that $|g_0(p, q)|^2 = \sum_{m,n \in \mathbb{Z}} \exp(4ik\pi(m - n)p) \exp\big(2i\pi\theta_{m,n}(q)\big)$ with

$$\theta_{m,n}(q) = (2kq + \mu)(m\tau - n\overline{\tau}) + k(m^2\tau - n^2\overline{\tau}) + \nu(n - m) + 2i\mu bq.$$

By exchanging the integrals and the sum and by applying Fubini's theorem, and since the integral of $\exp(4ik\pi(m - n)p)$ on $[0, 1]$ is equal to $\delta_{m,n}$, we obtain that

$$\|\phi_0\|_k^2 = 4\pi \sum_{m \in \mathbb{Z}} \underbrace{\int_0^1 \exp\big(-4\pi b(\mu(m + q) + k(m + q)^2)\big)\, dq}_{:=I_m}.$$

The change of variables $v = m + q$ yields

$$\sum_{m \in \mathbb{Z}} I_m = \sum_{m \in \mathbb{Z}} \int_m^{m+1} \exp\big(-4\pi b(\mu v + kv^2)\big)\, dv = \int_{\mathbb{R}} \exp\big(-4\pi b(\mu v + kv^2)\big)\, dv.$$

By forcing a square to appear in the exponential, we get that

$$\int_{\mathbb{R}} \exp\big(-4\pi b(\mu v + kv^2)\big)\, dv = \exp\left(\frac{\pi b\mu^2}{k}\right) \int_{\mathbb{R}} \exp\left(-4\pi bk\left(v + \frac{\mu}{2k}\right)^2\right)\, dv.$$

We conclude by using the change of variables $t = 2\sqrt{\pi bk}\left(v + (\mu/2k)\right)$ and the fact that the integral of $\exp(-t^2)$ over $\mathbb{R}$ is equal to $\sqrt{\pi}$. $\qquad\square$

Let us give explicit expressions for the elements ψ_ℓ of the orthonormal basis that we have obtained. We deduce from the previous lemma that

$$\psi_0(z) = \frac{(kb)^{1/4}\exp\left(-\pi b\mu^2/(2k)\right)}{\sqrt{2\pi}}\exp(2i\pi\mu z)$$
$$\times \left(\sum_{m\in\mathbb{Z}}\exp\left(2im\pi(2kz + \mu\tau + km\tau - v)\right)\right)t^k(z),$$

and by construction $\psi_\ell = (T^*_{f/2k})^\ell\psi_0 = T^*_{\ell f/2k}\psi_0$. On the one hand,

$$(T^*_{\ell f/2k}t^k)(z) = \exp\left(i\pi\ell\left(2z + \frac{\tau\ell}{2k}\right)\right)t^k(z),$$

while on the other hand

$$g_0\left(z + \frac{\ell}{2k}\right) = \exp\left(2i\pi\mu\left(z + \frac{\tau\ell}{2k}\right)\right)\sum_{m\in\mathbb{Z}}\exp\left(2im\pi(2kz + \mu\tau + km\tau - v + \tau\ell)\right).$$

Consequently, we obtain the following expression for ψ_ℓ, $0 \le \ell \le 2k - 1$:

$$\psi_\ell(z) = \frac{(kb)^{1/4}\exp\left(-\pi b\mu^2/(2k)\right)}{\sqrt{2\pi}}\exp(2i\pi\mu z)\left(\sum_{m\in\mathbb{Z}}\exp\left(2i\pi\theta_{\ell,m}(z)\right)\right)t^k(z)$$

where the function $\theta_{\ell,m}$ is defined as

$$\theta_{\ell,m}(z) = (\ell + 2km)\left(z + \frac{\mu\tau}{2k}\right) + (\ell + 2km)^2\frac{\tau}{4k} - mv.$$

We sum up the results so far.

Proposition 7.2.12. *The sections ψ_ℓ, $0 \le \ell \le 2k - 1$, defined in the above equation form an orthonormal basis of $\mathcal{H}_k^{\mu,\nu}$ satisfying*

$$\begin{cases} T^*_{e/2k}\psi_\ell = \exp(i\pi(\mu + \ell)/k)\psi_\ell, \\ T^*_{f/2k}\psi_\ell = \psi_{\ell+1} \end{cases}$$

for $\ell \in \mathbb{Z}/2k\mathbb{Z}$.

By Lemma 6.3.2, the Schwartz kernel of the projector Π_k satisfies

$$\Pi_k(z, w) = C_k \exp\big(2i\pi\mu(z - \overline{w})\big)$$

$$\times \left(\sum_{\ell=0}^{2k-1} \sum_{m,n \in \mathbb{Z}} \exp\Big(2i\pi \big(\theta_{\ell,m}(z) - \overline{\theta_{\ell,n}(w)} \big) \Big) \right) t^k(z) \otimes \overline{t}^k(w)$$

with $C_k = (2\pi)^{-1} \sqrt{kb} \exp\big(-\pi b \mu^2/k\big)$. We write $\theta_{\ell,m}(z) - \overline{\theta_{\ell,n}(w)} = \zeta_{\ell+2km, \ell+2kn}(z, w)$ where

$$\zeta_{r,s}(z, w) = r\left(z + \frac{\mu\tau}{2k} \right) + \frac{r^2\tau}{4k} - s\left(\overline{w} + \frac{\mu\overline{\tau}}{2k} \right) - \frac{s^2\overline{\tau}}{4k} + (s - r)\frac{\nu}{2k}.$$

We want to show that this kernel is as in Theorem 7.2.1, for the section E introduced in Example 7.1.9. In order to do so, we will need to evaluate its pointwise norm, away from the diagonal of $\mathbb{T}_\Lambda^2$ and near it. This is the purpose of the next two lemmas.

Lemma 7.2.13. *For every $\varepsilon \in (0, \frac{1}{2}]$, there exists a constant $C > 0$ such that for any two complex numbers $z = p_1 + \tau q_1$, $w = p_2 + \tau q_2$ satisfying $\mathrm{dist}(q_1 - q_2, \mathbb{Z}) \geq \varepsilon$, the inequality $\|\Pi_k(z, w)\| \leq C \exp(-k/C)$ holds.*

Proof. Let $z = p_1 + \tau q_1$ and $w = p_2 + \tau q_2$. We have that

$$\|\Pi_k(z, w)\| \leq \frac{\sqrt{kb}}{2\pi} \sum_{\ell=0}^{2k-1} \sum_{m,n \in \mathbb{Z}} \exp\big(-2\pi\kappa_{\ell+2km, \ell+2kn}(z, w)\big)$$

where $\kappa_{r,s}(z, w) = \frac{b\mu^2}{2k} + b\mu(q_1 + q_2) + \Im\big(\zeta_{r,s}(z, w)\big) + kb(q_1^2 + q_2^2)$. Moreover, for $r, s \in \mathbb{Z}$,

$$\Im\big(\zeta_{r,s}(z, w)\big) = rb\left(q_1 + \frac{\mu}{2k} \right) + \frac{r^2 b}{4k} + sb\left(q_2 + \frac{\mu}{2k} \right) + \frac{s^2 b}{4k},$$

hence we have that

$$\kappa_{r,s}(z, w) = b\left(k\left(q_1 + \frac{\mu}{2k} \right)^2 + r\left(q_1 + \frac{\mu}{2k} \right) + \frac{r^2}{4k} + k\left(q_2 + \frac{\mu}{2k} \right)^2 + s\left(q_2 + \frac{\mu}{2k} \right) + \frac{s^2}{4k} \right),$$

which can be written as

$$\kappa_{r,s}(z, w) = kb\left(\left(q_1 + \frac{r + \mu}{2k} \right)^2 + \left(q_2 + \frac{s + \mu}{2k} \right)^2 \right).$$

Thanks to the inequality $(x - y)^2 \leq 2(x^2 + y^2)$, valid for $x, y \in \mathbb{R}$, we obtain that there exists a constant $C > 0$ such that

$$\kappa_{r,s}(z, w) \geq Ckb\left(\left(q_1 + \frac{r + \mu}{2k} \right)^2 + \left(q_1 - q_2 + \frac{r - s}{2k} \right)^2 \right)$$

and it follows by setting $q = q_1 - q_2$ and replacing $m - n$ by n that

$$\|\Pi_k(z,w)\| \leq \frac{\sqrt{kb}}{2\pi} \sum_{\ell=0}^{2k-1} \sum_{m,n\in\mathbb{Z}} \exp\left(-2\pi Ckb\left(\left(q_1 + m + \frac{\ell+\mu}{2k}\right)^2 + (q+n)^2\right)\right).$$

Therefore $\|\Pi_k(z,w)\| \leq (2\pi)^{-1}\sqrt{kb}S_{k,1}S_{k,2}$ with

$$S_{k,1} = \sum_{\ell=0}^{2k-1} \sum_{m\in\mathbb{Z}} \exp\left(-C'k\left(q_1 + m + \frac{\ell+\mu}{2k}\right)^2\right), \quad S_{k,2} = \sum_{n\in\mathbb{Z}} \exp\left(-C'k(q+n)^2\right)$$

where $C' = 2\pi bC$. We claim that $S_{k,1} = O(k^{3/2})$, this estimate being uniform in z, w; in order to see this, one can compare the series appearing in $S_{k,1}$ with an integral. We also claim that $S_{k,2} \leq C'' \exp(-k/C'')$ for some $C'' > 0$ depending only on ε. Indeed, since $\mathrm{dist}(q, \mathbb{Z}) \geq \varepsilon$, we have that

$$S_{k,2} \leq 2\sum_{n=0}^{+\infty} \exp\left(-C'k(n+\varepsilon)^2\right) \leq 2\left(\exp(-C'k\varepsilon^2) + \sum_{n=1}^{+\infty} \exp\left(-C'kn^2\right)\right).$$

The claim follows from the fact that

$$\sum_{n=1}^{+\infty} \exp\left(-C'kn^2\right) \leq \sum_{n=1}^{+\infty} \exp\left(-C'kn\right) = \frac{\exp(-C'k)}{1 - \exp(-C'k)}.$$

These two claims allow us to conclude the proof. $\qquad\square$

Exercise 7.2.14. Prove the claim about $S_{k,1}$ in the proof above.

This lemma implies that the kernel of Π_k is a $O(k^{-\infty})$ outside the diagonal of $\mathbb{T}_A^2$. The next lemma deals with the near-diagonal behaviour of this kernel. We define a section R_k as

$$R_k(z,w)$$
$$= C_k \exp(2i\pi\mu(z - \overline{w})) \left(\sum_{\ell=0}^{2k-1} \sum_{\substack{m,n\in\mathbb{Z} \\ m\neq n}} \exp\left(2i\pi\left(\zeta_{\ell+2km,\ell+2kn}(z,w)\right)\right)\right) t^k(z) \otimes \overline{t}^k(w),$$

which means that we consider the same formula as the one defining Π_k except that the diagonal terms in the double sum have been removed.

Lemma 7.2.15. There exists $C > 0$ such that, for any $z = p_1 + \tau q_1$, $w = p_2 + \tau q_2 \in \mathbb{C}$ satisfying $|q_1 - q_2| \leq \frac{1}{2}$, the inequality $\|R_k(z,w)\| \leq C\exp(-k/C)$ holds.

Proof. Proceeding exactly as in the proof of the previous lemma, we obtain that $\|R_k(z,w)\| \leq (2\pi)^{-1}\sqrt{kb}S_{k,1}\tilde{s}_{k,2}$ for the same $S_{k,1}$ and with

$$\tilde{s}_{k,2} = \sum_{n \in \mathbb{Z} \setminus \{0\}} \exp\left(-C'k(q+n)^2\right) \leq 2 \sum_{n=0}^{+\infty} \exp\left(-C'k\left(n+\tfrac{1}{2}\right)^2\right),$$

where the inequality follows from the fact that $|q| \leq \frac{1}{2}$. From here we conclude as in the above proof. $\qquad\square$

This lemma implies that, for (z, w) sufficiently close to the diagonal,

$$\Pi_k(z, w) = C_k \exp\left(2i\pi\mu(z-\overline{w})\right)\left(\sum_{m \in \mathbb{Z}} \exp\left(2i\pi\zeta_{m,m}(z, w)\right)\right) t^k(z) \otimes \overline{t}^k(w) + O(k^{-\infty}).$$

One readily checks that this reads

$$\Pi_k(z, w) = \frac{\sqrt{kb}}{2\pi}\left(\sum_{m \in \mathbb{Z}} \exp\left(2i\pi(m+\mu)(z-\overline{w}) - \frac{\pi b(m+\mu)^2}{k}\right)\right) t^k(z) \otimes \overline{t}^k(w) + O(k^{-\infty}).$$

We can simplify further thanks to the following lemma.

Lemma 7.2.16. *We have that*

$$\sum_{m \in \mathbb{Z}} \exp\left(2i\pi(m+\mu)(z-\overline{w}) - \frac{\pi b(m+\mu)^2}{k}\right)$$
$$= \sqrt{\frac{k}{b}} \sum_{n \in \mathbb{Z}} \exp\left(2i\pi\mu n - \frac{k\pi\left(n - (z-\overline{w})\right)^2}{b}\right).$$

Proof. Let $f(t) = \exp(2i\pi(z-\overline{w})t - k^{-1}\pi bt^2)$ and $g(t) = f(t+\mu)$. Poisson's summation formula reads $\sum_{m \in \mathbb{Z}} g(m) = \sum_{n \in \mathbb{Z}} \hat{g}(2\pi n)$ where the Fourier transform is defined as

$$\hat{g}(\xi) = \int_{\mathbb{R}} \exp(-it\xi)g(t)\,dt.$$

With this convention, $\hat{g}(\xi) = \exp(i\mu\xi)\hat{f}(\xi)$. Moreover, $f(t) = \exp(2i\pi(z-\overline{w})t)h(t)$ where $h(t) = \exp(-k^{-1}\pi bt^2)$, hence $\hat{f}(\xi) = \hat{h}(\xi - 2\pi(z-\overline{w}))$. Since it is standard that

$$\hat{h}(\xi) = \sqrt{\frac{k}{b}} \exp\left(-\frac{k\xi^2}{4\pi b}\right),$$

we finally obtain that

$$\hat{g}(\xi) = \sqrt{\frac{k}{b}} \exp\left(i\mu\xi - \frac{k}{4\pi b}(\xi - 2\pi(z-\overline{w}))^2\right),$$

which yields the result. $\qquad\square$

Consequently, we finally obtain that for (z, w) close to the diagonal,

$$\Pi_k(z,w) = \frac{k}{2\pi}\left(\sum_{n\in\mathbb{Z}}\exp\left(2i\pi\mu n - \frac{k\pi(n-(z-\overline{w}))^2}{b}\right)\right)t^k(z)\otimes\overline{t}^k(w) + O(k^{-\infty}).$$

We need one last lemma, regarding the section

$$S_k(z,w) = \left(\sum_{n\in\mathbb{Z}\setminus\{0\}}\exp\left(2i\pi\mu n - \frac{k\pi(n-(z-\overline{w}))^2}{b}\right)\right)t^k(z)\otimes\overline{t}^k(w) + O(k^{-\infty}).$$

Lemma 7.2.17. *There exists $\varepsilon > 0$ and $C > 0$ such that $\|S_k(z,w)\| \leq C\exp(-k/C)$ for any $z = p_1 + \tau q_1$, $w = p_2 + \tau q_2$ satisfying $|p_1 - p_2| \leq \varepsilon$ and $|q_1 - q_2| \leq \varepsilon$.*

Proof. We clearly have that

$$\|S_k(z,w)\| \leq \sum_{n\in\mathbb{Z}\setminus\{0\}}\exp\left(\frac{k\pi}{b}\Re\left((n-(z-\overline{w}))^2\right) - 2k\pi b(q_1^2 + q_2^2)\right),$$

and a straightforward computation shows that the quantity in the exponential is equal to

$$-\frac{k\pi}{b}\left((n-(p_2-p_1)+a(q_2-q_1))^2 + b^2(q_1-q_2)^2\right).$$

Consequently, we obtain that

$$\|S_k(z,w)\| \leq \sum_{n\in\mathbb{Z}\setminus\{0\}}\exp\left(-\frac{k\pi}{b}(n-(p_2-p_1)+a(q_2-q_1))^2\right),$$

and we conclude as in the proof of Lemma 7.2.15. $\qquad\square$

After gathering all the previous lemmas, we finally obtain that

$$\Pi_k(z,w) = \frac{k}{2\pi}\exp\left(-\frac{k\pi(z-\overline{w})^2}{b}\right)t^k(z)\otimes\overline{t}(w) + O(k^{-\infty})$$

for (z,w) sufficiently close to the diagonal. This is consistent with Theorem 7.2.1, with E as in Example 7.1.9.

Additionally, we can now sketch the proof of the claim in Example 5.2.7. For $\lambda \in \Lambda$, the kernel K_k of $T^*_{\lambda/2k}$ is given by $K_k(\cdot,w) = T^*_{\lambda/2k}\Pi_k(\cdot,w)$. By writing $\lambda = p_\lambda e + q_\lambda f$ and $z_\lambda = p_\lambda + \tau q_\lambda$, we compute

$$\left(T^*_{\lambda/2k}t^k\right)(z) = \exp\left(2i\pi q_\lambda\left(z + \frac{z_\lambda}{2k}\right)\right)t^k(z).$$

Since moreover

$$\exp\left(-\frac{k\pi}{b}\left(z + \frac{z_\lambda}{2k} - \overline{w}\right)^2\right) = \exp\left(-\frac{k\pi(z-\overline{w})^2}{b}\right)\exp\left(-\frac{\pi z_\lambda}{b}\left(z - \overline{w} + \frac{z_\lambda}{4k}\right)\right),$$

we finally obtain thanks to the above formula for $\Pi_k(\cdot,\cdot)$ that

$$K_k(z,w) = \Pi_k(z,w)G_\lambda(z,w,k) + O\big(k^{-\infty}\big)$$

near the diagonal, where the function G_λ is defined as

$$G_\lambda(z,w,k) = \exp\left(2\mathrm{i}\pi q_\lambda\left(z + \frac{z_\lambda}{2k}\right) - \frac{\pi z_\lambda}{b}\left(z - \overline{w} + \frac{z_\lambda}{2k}\right)\right).$$

One readily checks that

$$G_\lambda(z,z,k) = \exp\left(2\mathrm{i}\pi(q_\lambda p - qp_\lambda) - \frac{\pi}{2bk}|z_\lambda|^2\right) = g_\lambda(z) + O\big(k^{-1}\big),$$

where we recall that $g_\lambda(x) = \exp(-\mathrm{i}\omega(\lambda,x)/2)$. We claim that this implies that $T^*_{\lambda/2k} = T_k(g_\lambda) + O\big(k^{-\infty}\big)$. One way to prove this is to check that the kernel $\widetilde{K}_k$ of $T_k(g_\lambda)$ is of the form

$$\widetilde{K}_k(z,w) = \Pi_k(z,w)\widetilde{G}_\lambda(z,w,k) + O\big(k^{-\infty}\big)$$

for some function $\widetilde{G}_\lambda(\cdot,\cdot,k)$ having an asymptotic expansion in non-positive powers of k whose first term coincides with g_λ on the diagonal. This is a general fact which can be proved by writing

$$\widetilde{K}_k(z,w) = \int_u \Pi_k(z,u)g_\lambda(u)\Pi_k(u,w)$$

and by applying the stationary phase lemma.

7.3 Idea of Proof of the Projector Asymptotics

In the rest of this section, we will very briefly sketch an idea of derivation of the asymptotics of the projector. This is a difficult result and writing a complete proof here would be too ambitious. The approach that we explain here is due to Berman, Berndtsson and Sjöstrand [6]. The main idea is that the Szegő projector is characterised by the fact that it is a reproducing kernel for $\mathcal{H}_k$. This means the following. Let $x \in M$ and let $u \in L_x$ be such that $h_x(u,u) = 1$. Then the formula $\xi^u_k(y) = \Pi_k(y,x)\cdot u^k$ defines an element of $\mathcal{H}_k$ (it is a coherent vector, see Chapter 9). It satisfies the reproducing property, that is, for any other element ϕ of $\mathcal{H}_k$, we have the equality

$$\phi(x) = \langle\phi, \xi^u_k\rangle_k u^k. \tag{7.2}$$

Indeed, we have that

$$\langle \phi, \xi_k^u \rangle_k = \int_M h_y^k \big(\phi(y), \xi_k^u(y) \big) \, \mu(y) = \int_M \overline{\xi_k^u(y)} \cdot \phi(y) \, \mu(y).$$

Moreover, we can write

$$\overline{\xi_k^u(y)} = \bar{u}^k \cdot \overline{\Pi_k(y, x)} = \bar{u}^k \cdot \Pi_k(x, y),$$

where the last equality follows from Proposition 6.3.6 since Π_k is self-adjoint. Consequently,

$$\langle \phi, \xi_k^u \rangle_k = \bar{u}^k \cdot \int_M \Pi_k(x, y) \cdot \phi(y) \, \mu(y) = \bar{u}^k \cdot (\Pi_k \phi)(x) = \bar{u}^k \cdot \phi(x)$$

because ϕ belongs to $\mathcal{H}_k$. This yields (7.2).

The proof is divided into two parts, as follows. Firstly, one can construct near each point of M a local section having the desired asymptotic expansion, and satisfying the same reproducing property up to some error; this section yields, in turn, a local section of $L \boxtimes \bar{L}$. Secondly, one can show that Π_k must agree with this local section up to some error.

Local Reproducing Kernels

Let U be an open subset of M endowed with a local non-vanishing holomorphic section s. Let $H = h(s, s)$ and let $\phi = -\log H$. Since $-i\omega$ is the curvature of the Chern connection on $L \to M$, we have that

$$\omega = i\bar{\partial}\partial(\log H) = i\partial\bar{\partial}\phi$$

on U. Any local holomorphic section of L^k is of the form $f s^k$, where f is a holomorphic function. For any two smooth local functions f, g, we define the quantity

$$\langle f, g \rangle_{\phi, k} = \int_U f\bar{g} \exp(-k\phi) \, \mu;$$

this defines a scalar product on $L^2(U, \exp(-k\phi)\mu)$. Observe that it is similar to the scalar product on $L^2(M, L^k)$. Let $\| \cdot \|_{\phi, k}$ be the associated norm. We define the space

$$\mathcal{H}_{\phi, k}(U) = \{ f \colon U \to \mathbb{C} \text{ holomorphic} \mid \|f\|_{\phi, k} < +\infty \}$$

of holomorphic functions on U with finite norm.

Let us consider local holomorphic coordinates, so that the coordinate open set is the unit ball B of $\mathbb{C}^n$. Fix a smooth function χ with compact support contained in B and equal to one on the ball of radius $\frac{1}{2}$. Let $K_k \colon U^2 \to \mathbb{C}$ be a local smooth function; we associate to K_k the local function

$$\zeta_k^x \colon U \to \mathbb{C}, \quad y \mapsto K_k(y, x).$$

Note that this construction is consistent with the one above; indeed, the vector $u = \exp(\phi(x)/2)s(x)$ is an element of L_x with unit norm. Such a function K_k is called a *reproducing kernel* modulo $O(k^{-N})$ for $\mathcal{H}_{\phi,k}(U)$ if for any local holomorphic function f,

$$f(x) = \langle \chi f, \zeta_k^x \rangle_{\phi,k} + O\big(k^{-N} \exp(k\phi(x)/2)\big) \|f\|_{\phi,k}$$

uniformly near the origin.

Let $\widetilde{\psi}$ be as in the proof of Proposition 7.1.2. Namely, $\widetilde{\psi}$ is holomorphic (respectively anti-holomorphic) in the left variable (respectively in the right variable) up to a flat function and satisfies $\widetilde{\psi}(x,x) = -i\phi(x)$.

Proposition 7.3.1 ([6, Proposition 2.7]). *There exist smooth functions $(b_\ell)_{\ell \geq 0}$ such that for every $N \geq 0$, there exists a reproducing kernel $K_k^{(N)}$ modulo $O\big(k^{n-N-1}\big)$ for $\mathcal{H}_{\phi,k}(U)$ such that*

$$K_k^{(N)}(x,y) = \left(\frac{k}{2\pi}\right)^n \exp\big(ik\widetilde{\psi}(x,y)\big)\big(b_0(x,y) + k^{-1}b_1(x,y) + \cdots + k^{-N}b_N(x,y)\big).$$

Roughly speaking, the idea is to investigate the behaviour of well-chosen contour integrals.

From Local to Global

Let $K_k^{(N)}$ be as in the above proposition. Then we define a local section of $L \boxtimes \bar{L} \to U^2$ by the formula

$$\widetilde{K_k^{(N)}}(x,y) = K_k^{(N)}(x,y)\, s^k(x) \otimes \bar{s}^k(y).$$

In particular, we have that

$$\widetilde{K_k^{(N)}}(x,y) = \left(\frac{k}{2\pi}\right)^n E^k(x,y)\big(b_0(x,y) + k^{-1}b_1(x,y) + \cdots + k^{-N}b_N(x,y)\big).$$

Therefore, the following result implies Theorem 7.2.1.

Theorem 7.3.2 ([6, Theorem 3.1]). *If $x, y \in U$ are close enough, then*

$$\Pi_k(x,y) = \widetilde{K_k^{(N)}}(x,y) + O\big(k^{n-N-1}\big).$$

The idea behind the proof of this theorem is to use the reproducing property to compare these two kernels.

Chapter 8
Proof of Product and Commutator Estimates

The aim of this chapter is to prove Theorems 5.2.2 and 5.2.3.

8.1 Corrected Berezin–Toeplitz Operators

Given a function $f \in \mathcal{C}^\infty(M, \mathbb{R})$, we introduce the corrected Berezin–Toeplitz quantisation of f:

$$T_k^c(f) = \Pi_k \left(f + \frac{1}{ik} \nabla_{X_f}^k \right) : \mathcal{H}_k \to \mathcal{H}_k \tag{8.1}$$

where X_f is the Hamiltonian vector field associated with f. The operator

$$P_k(f) = f + \frac{1}{ik} \nabla_{X_f}^k : \mathcal{C}^\infty(M, L^k) \to \mathcal{C}^\infty(M, L^k)$$

is called the *Kostant–Souriau operator* associated with f. The Kostant–Souriau operators satisfy the following nice properties.

Lemma 8.1.1. *For any* $f, g \in \mathcal{C}^\infty(M, \mathbb{R})$,

$$P_k(fg) = P_k(f)P_k(g) - \frac{1}{ik}\left(\{f, g\} + \frac{1}{ik} \nabla_{X_f}^k \nabla_{X_g}^k \right).$$

Proof. Since $X_{fg} = fX_g + gX_f$, we have that

$$P_k(fg) = f\left(g + \frac{1}{ik} \nabla_{X_g}^k \right) + g\left(\frac{1}{ik} \nabla_{X_f}^k \right) = fP_k(g) + g\left(\frac{1}{ik} \nabla_{X_f}^k \right).$$

We can rewrite this as

$$P_k(fg) = P_k(f)P_k(g) - \frac{1}{ik} \nabla_{X_f}^k P_k(g) + g\left(\frac{1}{ik} \nabla_{X_f}^k \right).$$

© Springer International Publishing AG, part of Springer Nature 2018
Y. Le Floch, *A Brief Introduction to Berezin–Toeplitz Operators on Compact Kähler Manifolds*, CRM Short Courses,
https://doi.org/10.1007/978-3-319-94682-5_8

Let us simplify the second term of the right-hand side; for $\phi \in \mathcal{C}^\infty(M, L^k)$, one has

$$\nabla^k_{X_f}(P_k(g)\phi) = (\mathcal{L}_{X_f}g)\phi + g\nabla^k_{X_f}\phi + \frac{1}{ik}\nabla^k_{X_f}\nabla^k_{X_g}\phi.$$

Using that $\mathcal{L}_{X_f}g = \{f, g\}$, this implies that

$$P_k(fg) = P_k(f)P_k(g) - \frac{1}{ik}\left(\{f,g\} + \frac{1}{ik}\nabla^k_{X_f}\nabla^k_{X_g}\right). \qquad \Box$$

This shows that $P_k(fg)$ differs from $P_k(f)P_k(g)$ by a remainder "of order k^{-1}". It turns out that for commutators, however, there is an exact (i.e. without remainder) correspondence principle for Kostant–Souriau operators.

Lemma 8.1.2. *For any $f, g \in \mathcal{C}^\infty(M, \mathbb{R})$,*

$$[P_k(f), P_k(g)] = \frac{1}{ik}P_k(\{f, g\}).$$

Proof. Since $P_k(gf) = P_k(fg)$, the previous lemma yields

$$[P_k(f), P_k(g)] = \frac{1}{ik}\left(\{f,g\} + \frac{1}{ik}\nabla^k_{X_f}\nabla^k_{X_g} - \{g,f\} - \frac{1}{ik}\nabla^k_{X_g}\nabla^k_{X_f}\right).$$

This can be rewritten as

$$[P_k(f), P_k(g)] = \frac{1}{ik}\left(2\{f,g\} + \frac{1}{ik}[\nabla^k_{X_f}, \nabla^k_{X_g}]\right).$$

Moreover, by definition of the curvature, we have that

$$[\nabla^k_{X_f}, \nabla^k_{X_g}] = \mathrm{curv}(\nabla^k)(X_f, X_g) + \nabla^k_{[X_f, X_g]},$$

which yields, since $\mathrm{curv}(\nabla^k) = -ik\omega$, and since $[X_f, X_g]$ is the Hamiltonian vector field associated with $\{f, g\}$,

$$[\nabla^k_{X_f}, \nabla^k_{X_g}] = -ik\{f, g\} + \nabla^k_{X_{\{f,g\}}}.$$

Putting all these equalities together, we finally obtain that

$$[P_k(f), P_k(g)] = \frac{1}{ik}\left(\{f,g\} + \frac{1}{ik}\nabla^k_{X_{\{f,g\}}}\right),$$

which was to be proved. $\Box$

The idea behind the proof of Theorems 5.2.2 and 5.2.3 is to derive from the properties above some estimates for the corrected Berezin–Toeplitz operators and to take profit of these estimates by comparing the corrected operator $T^c_k(f)$ with the usual Berezin–Toeplitz operator $T_k(f)$. In order to do so, we will need a result due to Tuynman [46], but let us first introduce some notation. Let $g = \omega(\cdot, j\cdot)$ be the Kähler metric on M, let μ_g be the associated volume form, let grad_g be the associated gradient, and let Δ be the associated Laplacian. We recall that for any $f \in \mathcal{C}^2(M)$,

$$\Delta f = \mathrm{div}_g(\mathrm{grad}_g f)$$

where the divergence $\mathrm{div}_g(X)$ of a vector field X on M is the function defined by the equality

$$\mathcal{L}_X \mu_g = \mathrm{div}_g(X)\mu_g.$$

Proposition 8.1.3 (Tuynman's lemma). *Let $X \in \mathcal{C}^1(M, TM \otimes \mathbb{C})$. Then*

$$\Pi_k \nabla_X^k \Pi_k = -\Pi_k \, \mathrm{div}_g(X^{1,0})\Pi_k,$$

where we recall that $X^{1,0} = (X - ijX)/2$. Furthermore, if $f \in \mathcal{C}^2(M, \mathbb{R})$, then

$$\Pi_k \left(\frac{1}{ik} \nabla_{X_f}^k \right) \Pi_k = -\frac{1}{2k} \Pi_k(\Delta f)\Pi_k.$$

The following corollary is immediate.

Corollary 8.1.4. *For every $X \in \mathcal{C}^1(M, TM \otimes \mathbb{C})$,*

$$\left\| \Pi_k \left(\frac{1}{ik} \nabla_X^k \right) \Pi_k \right\| = O(k^{-1})\|X\|_1.$$

In particular, for every $f \in \mathcal{C}^2(M, \mathbb{R})$,

$$\left\| \Pi_k \left(\frac{1}{ik} \nabla_{X_f}^k \right) \Pi_k \right\| = O(k^{-1})\|f\|_2.$$

Consequently, for every $f \in \mathcal{C}^2(M, \mathbb{R})$, $\|T_k^c(f) - T_k(f)\| = O(k^{-1})\|f\|_2$.

Proof of Proposition 8.1.3. Set $Y = X^{1,0}$. By virtue of Lemma 8.1.5 below, proving the first statement amounts to showing that for every $\phi \in \mathcal{H}_k$,

$$\langle \Pi_k(\nabla_X^k \phi), \phi \rangle_k = -\langle \Pi_k(\mathrm{div}_g(Y)\phi), \phi \rangle_k.$$

Using the facts that Π_k is self-adjoint and that $\Pi_k \phi = \phi$ whenever ϕ belongs to $\mathcal{H}_k$, we only need to prove that

$$\forall \phi \in \mathcal{H}_k, \quad \langle \nabla_X^k \phi, \phi \rangle_k = -\langle \mathrm{div}_g(Y)\phi, \phi \rangle_k. \tag{8.2}$$

Recall that $\mu_g = \mu$ the Liouville measure on M. We have that

$$\langle \mathrm{div}_g(Y)\phi, \phi \rangle_k = \int_M \mathrm{div}_g(Y) h_k(\phi, \phi) \, \mu_g = \int_M h_k(\phi, \phi) \, \mathcal{L}_Y \mu_g. \tag{8.3}$$

Now, by integrating the equality

$$\mathcal{L}_Y \big(h_k(\phi, \phi)\mu_g\big) = \mathcal{L}_Y \big(h_k(\phi, \phi)\big)\mu_g + h_k(\phi, \phi)\mathcal{L}_Y \mu_g,$$

we obtain that
$$\int_M h_k(\phi,\phi)\,\mathcal{L}_Y\mu_g = -\int_M \mathcal{L}_Y\big(h_k(\phi,\phi)\big)\mu_g.$$

Indeed, by Cartan's formula, and using the fact that $h_k(\phi,\phi)\mu_g$ is closed, we have that $\mathcal{L}_Y(h_k(\phi,\phi)\mu_g) = \mathrm{d}\big(i_Y(h_k(\phi,\phi)\mu_g)\big)$, thus its integral on M vanishes. Coming back to (8.3), this yields

$$\langle \mathrm{div}_g(Y)\phi,\phi\rangle_k = -\int_M \mathcal{L}_Y\big(h_k(\phi,\phi)\big)\mu_g = -\int_M \Big(h_k\big(\nabla^k_Y\phi,\phi\big) + h_k\big(\phi,\nabla^k_{\overline{Y}}\phi\big)\Big)\mu_g,$$

where the second equality comes from the fact that ∇^k and h_k are compatible. But $\overline{Y}$ is a section of $T^{0,1}M$, and ϕ is a holomorphic section of L^k, so $\nabla^k_{\overline{Y}}\phi = 0$, which implies that $\nabla^k_X\phi = \nabla^k_Y\phi$ since $X = Y + \overline{Y}$, and (8.2) is proved.

We now want to apply this to X_f where f belongs to $\mathcal{C}^2(M,\mathbb{R})$. Observe that

$$\mathrm{div}_g\big(X_f^{1,0}\big) = \tfrac{1}{2}\big(\mathrm{div}_g(X_f) - \mathrm{i}\,\mathrm{div}_g(jX_f)\big).$$

We claim that $\mathrm{div}_g(X_f) = 0$; indeed, since $\mu_g = \mu$, we have that

$$\mathrm{div}_g(X_f)\mu_g = \mathcal{L}_{X_f}\mu_g = \mathcal{L}_{X_f}\mu = 0.$$

Consequently, $\mathrm{div}_g(X_f^{1,0}) = -(\mathrm{i}/2)\,\mathrm{div}_g(jX_f)$. Thanks to Lemma 2.6.1, this yields

$$\mathrm{div}_g(X_f^{1,0}) = \frac{\mathrm{i}}{2}\,\mathrm{div}(\mathrm{grad}_g f) = \frac{\mathrm{i}}{2}\Delta f,$$

and the second statement follows. $\qquad\square$

Lemma 8.1.5. *Let T be a bounded operator acting on a complex Hilbert space $\mathcal{H}$. If $\langle T\xi,\xi\rangle = 0$ for every $\xi \in \mathcal{H}$, then $T = 0$.*

Proof. This is a standard exercise but we still prove it. Let $\xi,\eta \in \mathcal{H}$. Then

$$0 = \langle T(\xi+\eta),\xi+\eta\rangle = \langle T\xi,\xi\rangle + \langle T\xi,\eta\rangle + \langle T\eta,\xi\rangle + \langle T\eta,\eta\rangle$$

which yields

$$\langle T\xi,\eta\rangle = -\langle T\eta,\xi\rangle.$$

Replacing η by $\mathrm{i}\eta$, this implies that

$$-\mathrm{i}\langle T\xi,\eta\rangle = -\mathrm{i}\langle T\eta,\xi\rangle,$$

and combining these two equalities yields $\langle T\xi,\eta\rangle = 0$. $\qquad\square$

8.2 Unitary Evolution of Kostant–Souriau Operators

The goal of this section is to give an alternate, more geometric proof of Lemma 8.1.2, and to use this as an excuse to address the topic of the Schrödinger equation for these operators. More precisely, given a function $f \in C^\infty(M, \mathbb{R})$, we want to look for solutions of

$$\frac{\mathrm{d}\Psi_t}{\mathrm{d}t} = -\mathrm{i}kP_k(f)\Psi_t, \quad t \in \mathbb{R}, \tag{8.4}$$

where Ψ_t is a smooth section of $L^k \to M$ and $\Psi_0 \in C^\infty(M, L^k)$ is a given initial condition. We can solve this equation as follows. Given a path $\gamma: [0, T] \to M$, let

$$T_\gamma^k: L_{\gamma(0)}^k \to L_{\gamma(T)}^k$$

be the parallel transport operator in L^k with respect to ∇^k. Moreover, let ϕ^t be the Hamiltonian flow of f at time t.

Proposition 8.2.1. *Given $\Psi_0 \in C^\infty(M, L^k)$, the family of sections $\Psi_t \in C^\infty(M, L^k)$ defined as*

$$\Psi_t\big(\phi^t(m)\big) = \exp\big(-\mathrm{i}ktf(m)\big)T_{(\phi^s(m))_{s \in [0,t]}}^k\big(\Psi_0(m)\big)$$

for every $m \in M$, is a solution of (8.4) with initial condition Ψ_0.

This defines an operator $U_k(t): C^\infty(M, L^k) \to C^\infty(M, L^k)$ sending Ψ_0 to Ψ_t, which describes the prequantum evolution of the system.

Proof. We fix $m \in M$ and $\Psi_0 \in C^\infty(M, L^k)$. We claim that it is enough to prove the proposition for t so small that for every $s \in [-t, t]$, the point $\phi^s(m)$ belongs to a trivialisation open set V for L. This is because the operator $U_k(t)$ satisfies the semigroup relation $U_k(t_1 + t_2) = U_k(t_2)U_k(t_1)$.

Let u be a local non-vanishing section of L over V, and let $\varphi = hu^k$ for some $h \in C^\infty(V, \mathbb{R})$. Moreover, let α be the differential form such that $\nabla s = -\mathrm{i}\alpha \otimes s$. Then we can write $P_k(f)\varphi = (\widetilde{P}_k(f)h)u^k$ with

$$\widetilde{P}_k(f)h = (f - i_{X_f}\alpha)h + \frac{1}{\mathrm{i}k}\mathcal{L}_{X_f}h. \tag{8.5}$$

Moreover, a standard computation yields

$$T_{(\phi^s(m))_{s \in [0,t]}}^k\big(\varphi(m)\big) = \exp\left(\mathrm{i}k\int_0^t (\phi^s)^*(i_{X_f}\alpha)\,\mathrm{d}s\right)h(m)u^k\big(\phi^t(m)\big),$$

and consequently, if $\Psi_0 = h_0 s^k$ on V, then $\Psi_t = h_t s^k$ on V where

$$h_t(m) = \exp\left(\mathrm{i}k\left(\int_{-t}^0 (\phi^s)^*(i_{X_f}\alpha)\,\mathrm{d}s - tf(m)\right)\right)h_0\big(\phi^{-t}(m)\big).$$

for every $m \in M$. We only need to compare the time derivative of h_t and $\widetilde{P}_k(f)h_t$. To simplify notation, we will write

$$\theta(t, m) = \int_{-t}^{0} (\phi^s)^*(i_{X_f}\alpha)(m)\,\mathrm{d}s - tf(m).$$

On the one hand,

$$\frac{\mathrm{d}h_t}{\mathrm{d}t} = \exp\big(ik\theta(t, \cdot)\big)\big(-(\phi^{-t})^*(\mathcal{L}_{X_f}h_0) + ik\big((\phi^{-t})^*(i_{X_f}\alpha) - f\big)(\phi^{-t})^*h_0\big).$$

On the other hand, we have that

$$\mathcal{L}_{X_f}h_t = \exp\big(ik\theta(t, \cdot)\big)\bigg((\phi^{-t})^*(\mathcal{L}_{X_f}h_0) + ik\big((\phi^{-t})^*h_0\big)\int_{-t}^{0}\mathcal{L}_{X_f}\big((\phi^s)^*(i_{X_f}\alpha)\big)\,\mathrm{d}s\bigg).$$

Using Cartan's formula, we have that

$$\mathrm{d}\big((\phi^s)^*(i_{X_f}\alpha)\big) = (\phi^s)^*\big(\mathrm{d}(i_{X_f}\alpha)\big) = (\phi^s)^*(\mathcal{L}_{X_f}\alpha) - (\phi^s)^*(i_{X_f}\,\mathrm{d}\alpha).$$

Since $\mathrm{d}\alpha = i\,\mathrm{curv}(L) = \omega$ and $(\phi^s)^*(\mathcal{L}_{X_f}\alpha) = \mathrm{d}(\phi^s)^*\alpha/\mathrm{d}s$, we can write

$$\int_{-t}^{0}\mathcal{L}_{X_f}\big((\phi^s)^*(i_{X_f}\alpha)\big)\,\mathrm{d}s = i_{X_f}\alpha - (\phi^{-t})^*(i_{X_f}\alpha),$$

therefore we finally obtain that

$$\widetilde{P}_k(f)h_t = \exp\big(ik\theta(t, \cdot)\big)\bigg(\big(f - (\phi^{-t})^*(i_{X_f}\alpha)\big)(\phi^{-t})^*h_0 + \frac{1}{ik}(\phi^{-t})^*(\mathcal{L}_{X_f}h_0)\bigg),$$

which yields the desired formula $-ik\widetilde{P}_k(f)h_t = (\mathrm{d}h_t/\mathrm{d}t)$.   □

One can check that $U_k(t)$ extends to a unitary operator on $L^2(M, L^k)$. It turns out that the Kostant–Souriau operators satisfy an exact version of Egorov's theorem (Theorem 5.3.2).

Proposition 8.2.2. *Let $f \in \mathcal{C}^\infty(M, \mathbb{R})$ and let $U_k(t)$ be the evolution operator associated with $P_k(f)$. Then*

$$U_k(t)^* P_k(g) U_k(t) = P_k(g \circ \phi^t)$$

for every $g \in \mathcal{C}^\infty(M, \mathbb{R})$, where ϕ^t is the Hamiltonian flow of f at time t.

Proof. Again, we can work in a trivialisation open set for L, since

$$U_k(t_1 + t_2)^* P_k(g) U_k(t_1 + t_2) = U_k(t_2)^* U_k(t_1)^* P_k(g) U_k(t_1) U_k(t_2),$$
$$g \circ \phi^{t_1 + t_2} = g \circ \phi^{t_1} \circ \phi^{t_2}.$$

Hence we keep the same notation as in the proof of the previous proposition. If $U_k(t)\Psi_0 = h_t u^k$ on V, the computations performed in this proof yield

$$\mathrm{d}h_t = \exp\big(ik\theta(t,\cdot)\big)$$

$$\times\left((\phi^{-t})^*(\mathrm{d}h_0) + ik\left(\alpha - (\phi^{-t})^*\alpha - \int_{-t}^0 (\phi^s)^*(i_{X_f}\,\mathrm{d}\alpha)\,\mathrm{d}s - t\,\mathrm{d}f\right)(\phi^{-t})^*h_0\right).$$

We can simplify this further because

$$(\phi^s)^*(i_{X_f}\,\mathrm{d}\alpha) = (\phi^s)^*(i_{X_f}\omega) = -(\phi^s)^*(\mathrm{d}f) = -\mathrm{d}\big((\phi^s)^*f\big) = -\mathrm{d}f,$$

hence we obtain that

$$\mathcal{L}_{X_g}h_t = \exp\big(ik\theta(t,\cdot)\big)\left((\phi^{-t})^*(\mathcal{L}_{X_g}h_0) + ik\left(i_{X_g}\alpha - i_{X_g}\big((\phi^{-t})^*\alpha\big)\right)(\phi^{-t})^*h_0\right).$$

Therefore, (8.5) yields

$$\widetilde{P}_k(g)h_t = \exp\big(ik\theta(t,\cdot)\big)\left(\frac{1}{ik}(\phi^{-t})^*(\mathcal{L}_{X_g}h_0) + \left(g - i_{X_g}\big((\phi^{-t})^*\alpha\big)\right)(\phi^{-t})^*h_0\right).$$

Consequently, if $U_k(t)^*P_k(g)U_k(t) = q_t u^k$ on V, we finally obtain that

$$q_t = \frac{1}{ik}\mathcal{L}_{X_{g\circ\phi^t}}h_0 + \big(g\circ\phi^t - i_{X_{g\circ\phi^t}}\alpha\big)h_0 = \widetilde{P}_k(g\circ\phi^t)h_0. \qquad \square$$

In order to reprove Lemma 8.1.2 with the help of these two results, it suffices to write the time derivative of $\phi_k(t) = U_k(t)^*P_k(g)U_k(t)\Psi_0$, for $\Psi_0 \in C^\infty(M, L^k)$, in two different ways. On the one hand, by definition of U_k,

$$\left.\frac{\mathrm{d}\phi_k}{\mathrm{d}t}\right|_{t=0} = ik[P_k(f), P_k(g)]\Psi_0.$$

On the other hand, since $\phi_k(t) = P_k(g\circ\phi^t)\Psi_0$, Lemma 5.3.3 implies that

$$\left.\frac{\mathrm{d}\phi_k}{\mathrm{d}t}\right|_{t=0} = P_k(\{f,g\})\Psi_0,$$

and we conclude by comparing these two equalities that the Kostant–Souriau operators satisfy the exact correspondence principle.

8.3 Product Estimate

We will need the following result, of which we will give a proof in Section 8.5.

Theorem 8.3.1. *There exists $C > 0$ such that for every $f \in C^2(M, \mathbb{R})$,*

$$\|[P_k(f), \Pi_k]\| \le Ck^{-1}\|f\|_2.$$

This estimate is fundamental and allows us to obtain product and commutator estimates. We now use it to prove Theorem 5.2.2. We compute the difference

$$T_k(f)T_k(g) - T_k(fg) = \Pi_k f[\Pi_k, g]\Pi_k = \Pi_k f[\Pi_k, P_k(g)]\Pi_k - \Pi_k f \left[\Pi_k, \frac{1}{ik}\nabla^k_{X_g}\right]\Pi_k.$$

Thanks to Theorem 8.3.1, we know that $\|\Pi_k f[\Pi_k, P_k(g)]\Pi_k\| = O(k^{-1})\|f\|_0\|g\|_2$. The other term can be estimated by writing it as

$$\Pi_k f \left[\Pi_k, \frac{1}{ik}\nabla^k_{X_g}\right]\Pi_k = \Pi_k f \Pi_k \left(\frac{1}{ik}\nabla^k_{X_g}\right)\Pi_k - \Pi_k \left(\frac{1}{ik}\nabla^k_{fX_g}\right)\Pi_k.$$

Both terms can be estimated using Corollary 8.1.4. The first one satisfies

$$\left\|\Pi_k f \Pi_k \left(\frac{1}{ik}\nabla^k_{X_g}\right)\Pi_k\right\| = O(k^{-1})\|f\|_0\|g\|_2,$$

whereas the second one satisfies

$$\left\|\Pi_k \left(\frac{1}{ik}\nabla^k_{fX_g}\right)\Pi_k\right\| = O(k^{-1})\|fX_g\|_1 = O(k^{-1})(\|f\|_0\|g\|_2 + \|f\|_1\|g\|_1).$$

This proves the first estimate of the theorem. To derive the second one, observe that $T_k(fg)$ is self-adjoint and that the adjoint of $T_k(f)T_k(g)$ is $T_k(g)T_k(f)$, and use the fact that the operator norm of the adjoint of an operator is the same as the norm of the operator.

8.4 Commutator Estimate

We first prove commutator estimates for corrected Berezin–Toeplitz operators.

Proposition 8.4.1. *For any* $f, g \in \mathcal{C}^2(M, \mathbb{R})$,

$$\left\|[T^c_k(f), T^c_k(g)] - \frac{1}{ik}T^c_k(\{f, g\})\right\| = O(k^{-2})\|f\|_2\|g\|_2.$$

Proof. We will compare $[T^c_k(f), T^c_k(g)]$ with $[P_k(f), P_k(g)]$. In order to do so, we compute:

$$\Pi_k[\Pi_k, P_k(f)][\Pi_k, P_k(g)]\Pi_k = \Pi_k P_k(f)[\Pi_k, P_k(g)]\Pi_k - \Pi_k P_k(f)\Pi_k[\Pi_k, P_k(g)]\Pi_k.$$

Expanding the first term on the right-hand side of this equality, we get

$$\Pi_k P_k(f)[\Pi_k, P_k(g)]\Pi_k = \Pi_k P_k(f)\Pi_k P_k(g)\Pi_k - \Pi_k P_k(f)P_k(g)\Pi_k$$

and the second term satisfies

$$\Pi_k P_k(f) \Pi_k [\Pi_k, P_k(g)] \Pi_k = \Pi_k P_k(f) \Pi_k P_k(g) \Pi_k - \Pi_k P_k(f) \Pi_k P_k(g) \Pi_k = 0.$$

Therefore, we have that

$$\Pi_k [\Pi_k, P_k(f)][\Pi_k, P_k(g)] \Pi_k = T_k^c(f) T_k^c(g) - \Pi_k P_k(f) P_k(g) \Pi_k.$$

Thanks to Theorem 8.3.1, the left-hand side is a $\mathcal{O}(k^{-2}) \|f\|_2 \|g\|_2$, thus

$$[T_k^c(f), T_k^c(g)] = \Pi_k [P_k(f), P_k(g)] \Pi_k + \mathcal{O}(k^{-2}) \|f\|_2 \|g\|_2$$

which yields, using Lemma 8.1.2,

$$[T_k^c(f), T_k^c(g)] = \frac{1}{ik} T_k^c(\{f, g\}) + \mathcal{O}(k^{-2}) \|f\|_2 \|g\|_2. \qquad \square$$

We now prove Theorem 5.2.3. Thanks to Proposition 8.1.3, we have that

$$T_k(f) = T_k^c(f) + \frac{1}{2k} T_k(\Delta f),$$

and similarly for g. Consequently, $[T_k(f), T_k(g)] = [T_k^c(f), T_k^c(g)] + R_k$, with

$$R_k = \frac{1}{2k} [T_k(\Delta f), T_k^c(g)] + \frac{1}{2k} [T_k^c(f), T_k(\Delta g)] + \frac{1}{4k^2} [T_k(\Delta f), T_k(\Delta g)].$$

Let us estimate R_k. Firstly, we have that

$$[T_k(\Delta f), T_k^c(g)] = [T_k(\Delta f), T_k(g)] - \frac{1}{2k} [T_k(\Delta f), T_k(\Delta g)].$$

Applying Theorem 5.2.2 to $\Delta f \in \mathcal{C}^1(M, \mathbb{R})$ and $g \in \mathcal{C}^3(M, \mathbb{R})$, we obtain that

$$[T_k(\Delta f), T_k(g)] = O(k^{-1})(\|f\|_2 \|g\|_2 + \|f\|_3 \|g\|_1) = O(k^{-1}) \|f, g\|_{1,3}.$$

Moreover, Lemma 5.1.2 implies that

$$[T_k(\Delta f), T_k(\Delta g)] = O(1) \|\Delta f\|_0 \|\Delta g\|_0 = O(1) \|f\|_2 \|g\|_2. \qquad (8.6)$$

It follows from these estimates that

$$\frac{1}{2k} [T_k(\Delta f), T_k^c(g)] = O(k^{-2}) \|f, g\|_{1,3}.$$

A similar reasoning leads to

$$\frac{1}{2k} [T_k^c(f), T_k(\Delta g)] = O(k^{-2}) \|f, g\|_{1,3}.$$

These two results combined with (8.6) imply that $R_k = O(k^{-2}) \|f, g\|_{1,3}$. Now, thanks to the previous proposition, we have that

$$[T_k^c(f), T_k^c(g)] = \frac{1}{ik} T_k^c(\{f,g\}) + O(k^{-2})\|f,g\|_{1,3}.$$

Therefore,

$$[T_k(f), T_k(g)] = \frac{1}{ik} T_k(\{f,g\}) + \frac{i}{2k^2} T_k(\Delta\{f,g\}) + O(k^{-2})\|f,g\|_{1,3},$$

and we conclude thanks to the estimate

$$T_k(\Delta\{f,g\}) = O(1)\|\Delta\{f,g\}\|_0 = O(1)\|f,g\|_{1,3},$$

which follows from Lemma 5.1.2.

8.5 Fundamental Estimates

This section, which follows the same lines as in the article [20], is devoted to the proof of Theorem 8.3.1; this strongly relies on the asymptotic expansion of the Schwartz kernel of the projector given by Theorem 7.2.1. Let $E \in \mathcal{C}^\infty(M \times \overline{M}, L^k \boxtimes \overline{L}^k)$ be as in this theorem, that is, satisfying the properties stated in Proposition 7.1.1. Let $U \subset M^2$ be the open set where E does not vanish; observe that U contains the diagonal Δ_M of M^2. Define as before a function $\varphi_E \in \mathcal{C}^\infty(U)$ and a differential form $\alpha_E \in \Omega^1(U) \otimes \mathbb{C}$ by the formulas

$$\varphi_E = -2\log\|E\|, \quad \widetilde{\nabla} E = -i\alpha_E \otimes E,$$

where we recall that $\widetilde{\nabla}$ is the connection induced by ∇ on $L \boxtimes \overline{L}$. The function φ_E vanishes along Δ_M and is positive outside Δ_M. We derived the following properties of φ_E and α_E in Lemmas 7.1.3 and 7.1.4:

(1) α_E vanishes along Δ_M,
(2) φ_E vanishes to second order along Δ_M,
(3) for every $x \in M$, the kernel of the Hessian of φ_E at (x,x) is equal to $T_{(x,x)}\Delta_M$, and this Hessian is positive definite on the complement of $T_{(x,x)}\Delta_M$.

In what follows, we will need the following additional property.

Lemma 8.5.1. *Let $f \in \mathcal{C}^2(M, \mathbb{R})$, and let $g \in \mathcal{C}^2(U, \mathbb{R})$ be defined by the formula $g(x,y) = f(x) - f(y)$. Then the function*

$$u = g - \alpha_E(X_f, X_f)$$

vanishes to second order along Δ_M.

Proof. It is clear that u vanishes along Δ_M since g and α_E do. Now, let Y and Z be two vector fields on M; we compute for $(y,z) \in U$

$$(\mathcal{L}_{(Y,Z)}u)(y,z) = (\mathcal{L}_Y f)(y) - (\mathcal{L}_Z f)(z) - \mathcal{L}_{(Y,Z)}(\alpha_E(X_f, X_f))(y,z).$$

As before, set $\widetilde{\omega} = p_1^*\omega - p_2^*\omega$ with p_1, p_2 the natural projections $M^2 \to M$. Therefore the first two terms in the above equation satisfy

$$\big(\mathcal{L}_Y f\big)(y) - \big(\mathcal{L}_Z f\big)(z) = \widetilde{\omega}\big((Y, Z), (X_f, X_f)\big)(x, y).$$

Moreover, since $\mathrm{d}\alpha_E = \mathrm{i}\,\mathrm{curv}(\widetilde{\nabla}) = \widetilde{\omega}$, the last term in the previous equation can be written as

$$\mathcal{L}_{(Y,Z)}\big(\alpha_E(X_f, X_f)\big) = \widetilde{\omega}\big((Y, Z), (X_f, X_f)\big) + \alpha_E([(Y, Z), (X_f, X_f)])$$
$$+ \mathcal{L}_{(X_f, X_f)}\big(\alpha_E(Y, Z)\big).$$

Thus we finally obtain that

$$\mathcal{L}_{(Y,Z)} u = \alpha_E([(X_f, X_f), (Y, Z)]) - \mathcal{L}_{(X_f, X_f)}\big(\alpha_E(Y, Z)\big).$$

The first term vanishes along Δ_M because α_E does. The second term vanishes along Δ_M because α_E vanishes along Δ_M and (X_f, X_f) is tangent to Δ_M. $\square$

These properties yield the following result. For $u \in \mathcal{C}^0(M^2, \mathbb{R})$, let $Q_k(u)$ be the operator acting on $\mathcal{C}^0(M, L^k)$ with Schwartz kernel $F_k(u) = \big(k/(2\pi)\big)^n E^k u$.

Lemma 8.5.2. *Taking a smaller U still containing Δ_M if necessary, for every compact subset $K \subset U$ and for every $p \in \mathbb{N}$, there exists a constant $C_{K,p} > 0$ such that for any $u \in \mathcal{C}^0(M^2, \mathbb{R})$ with support contained in K, and for every $k \geq 1$,*

$$\|Q_k(u)\| \leq C_{K,p}|u|_{K,p}k^{-p/2}$$

where $|u|_{K,p}$ is the supremum of $|u|\varphi_E^{-p/2}$ on $K \setminus \Delta_M$, which may be infinite.

Proof. Assume first that $K \subset V^2$, where $V \subset M$ is a trivialisation open set for M, with coordinates $x_1, \ldots, x_{2n}$, such that $V^2 \subset U$. So we may identify V with a subset of $\mathbb{R}^{2n}$ and assume that we are working in a subset of $\mathbb{R}^{4n}$. Since φ_E vanishes to second order along Δ_M, Taylor's formula with integral remainder yields

$$\varphi_E(x, y) = \frac{1}{2}\,\mathrm{d}^2\varphi_E(x, x)(v, v) + \int_0^1 \frac{(1-t)^2}{2}\,\mathrm{d}^3\varphi_E\big((1-t)(x, x) + t(x, y)\big)(v, v, v)\,\mathrm{d}t$$

with $v = (0, y - x)$. The last term is a $O\big(|x - y|^3\big)$ uniformly on K. Since $\mathrm{d}^2\varphi_E(x, x)$ is positive definite on the orthogonal of $\{x = y\} \subset \mathbb{R}^{4n}$, we have that

$$\lambda_{\min}(x)\|v\|^2 \leq \mathrm{d}^2\varphi_E(x, x)(v, v) \leq \lambda_{\max}(x)\|v\|^2$$

whenever $y \neq x$, where $\lambda_{\min}(x)$ (respectively $\lambda_{\max}(x)$) is the smallest (respectively largest) positive eigenvalue of $\mathrm{d}^2\varphi_E(x, x)$. Therefore, there exists $C > 0$ such that

$$\frac{\|x - y\|^2}{C} \leq \varphi_E(x, y) \leq C\|x - y\|^2 \tag{8.7}$$

for every $(x, y) \in K$. Now, let $u \in \mathcal{C}^0(M^2, \mathbb{R})$ be compactly supported in K. The previous estimate shows that for every $(x, y) \in K$, $x \neq y$,

$$\frac{|u(x, y)|}{\varphi_E(x, y)^{p/2}} \geq \frac{|u(x, y)|}{C^{p/2}\|x - y\|^p},$$

thus $|u(x, y)| \leq C^{p/2}|u|_{K,p}\|x - y\|^p$ on V^2. If $|u|_{K,p}$ is infinite, the result is obvious. If not, since $\|E\| = \exp(-\varphi_E/2)$, we have that

$$\int_M \|F_k(u)(x, y)\| \, \mathrm{d}x \leq \left(\frac{k}{2\pi}\right)^n C^{p/2}|u|_{K,p} \int_V \exp\left(-\frac{k\|x - y\|^2}{2C}\right)\|x - y\|^p \, \mathrm{d}x.$$

The integral on V is smaller that the integral on $\mathbb{R}^{2n}$ of the same integrand. The change of variable $v = \sqrt{k/C}(x - y)$ yields

$$\int_M \|F_k(u)(x, y)\| \, \mathrm{d}x \leq \frac{C^{p+n}}{(2\pi)^n} k^{-p/2}|u|_{K,p} \int_{\mathbb{R}^{2n}} \exp\left(-\frac{\|v\|^2}{2}\right)\|v\|^p \, \mathrm{d}v,$$

which implies that $\int_M \|F_k(u)(x, y)\| \, \mathrm{d}x \leq C_{K,p}^1 k^{-p/2}|u|_{K,p}$. A similar computation leads to $\int_M \|F_k(u)(x, y)\| \, \mathrm{d}y \leq C_{K,p}^2 k^{-p/2}|u|_{K,p}$ for some $C_{K,p}^2 > 0$. It follows from the Schur test that

$$\|Q_k(u)\| \leq C_{K,p} k^{-p/2}|u|_{K,p}$$

for some $C_{K,p} > 0$.

Let us now turn to the general case. Taking a smaller U, still containing the diagonal, if necessary, let $(V_i)_{1 \leq i \leq d}$ be a finite family of trivialisation sets of M such that $K \subset \bigcup_{i=1}^d V_i^2 \subset U$. Choose a partition of unity $\eta, (\eta_i)_{1 \leq i \leq d}$ subordinate to the open cover $M^2 \subset (M^2 \setminus K) \cup (\bigcup_{i=1}^d V_i^2)$. Let $u \in \mathcal{C}^0(M^2, \mathbb{R})$ be compactly supported in K; then

$$u = \sum_{i=1}^d \eta_i u, \quad Q_k(u) = \sum_{i=1}^d Q_k(\eta_i u).$$

It follows from the first part of the proof that

$$\|Q_k(\eta_i u)\| \leq C_{K,p,i} k^{-p/2}|\eta_i u|_{K,p} \leq C_{K,p,i} k^{-p/2}|u|_{K,p}$$

for some constants $C_{K,p,i} > 0$. We conclude by applying the triangle inequality. $\square$

Proposition 8.5.3. *For every $p \in \mathbb{N}$, for every $u \in \mathcal{C}^\infty(M^2, \mathbb{R})$ supported in U and vanishing to order p along Δ_M, there exists $C_u > 0$ such that for every $f \in \mathcal{C}^2(M, \mathbb{R})$,*

$$\|Q_k(u)\| \leq C_u k^{-p/2}, \quad \|[P_k(f), Q_k(u)]\| \leq C_u k^{-p/2-1}\|f\|_2,$$

where $P_k(f) = f + (1/(ik))\nabla_{X_f}^k : \mathcal{C}^\infty(M, L^k) \to \mathcal{C}^\infty(M, L^k)$ is the Kostant–Souriau operator associated with f.

Before proving this result, let us state several lemmas.

Lemma 8.5.4. *Let* $u \in \mathcal{C}^\infty(M^2, \mathbb{R})$ *be compactly supported in* U, *and let* $f \in \mathcal{C}^2(M, \mathbb{R})$. *Let* $g \in \mathcal{C}^2(M^2, \mathbb{R})$ *be defined by the formula* $g(x, y) = f(x) - f(y)$ *as before, and define the vector field* $Y_f = (X_f, X_f)$ *on* M^2. *Then*

$$[P_k(f), Q_k(u)] = Q_k\Big((g - \alpha_E(Y_f))u\Big) + \frac{1}{ik}Q_k(\mathcal{L}_{Y_f}u).$$

Proof. We start by writing

$$[P_k(f), Q_k(u)] = fQ_k(u) - Q_k(u)f + \frac{1}{ik}\Big(\nabla^k_{X_f} \circ Q_k(u) - Q_k(u) \circ \nabla^k_{X_f}\Big).$$

The Schwartz kernel of $fQ_k(u) - Q_k(u)f$ is equal to $f(x)F_k(u)(x, y) - F_k(u)(x, y)f(y)$. By Lemma 6.4.3, the Schwartz kernel of $\nabla^k_{X_f} \circ Q_k(u)$ is equal to $(\nabla^k_{X_f} \boxtimes \mathrm{id})F_k(u)$. By Lemma 6.4.4, the Schwartz kernel of $Q_k(u) \circ \nabla^k_{X_f}$ is equal to $-(\mathrm{id} \boxtimes \nabla^k_{X_f})F_k(u)$ since $\mathrm{div}(X_f) = 0$. Therefore, the Schwartz kernel of $[P_k(f), Q_k(u)]$ is given by

$$\left(f \boxtimes \mathrm{id} - \mathrm{id} \boxtimes f + \frac{1}{ik}\widetilde{\nabla}^k_{(X_f, X_f)}\right)F_k(u).$$

Remembering the definition of α_E, and since u has support in U, we have that

$$\widetilde{\nabla}^k_{(X_f, X_f)}(E^k u) = u\widetilde{\nabla}^k_{Y_f}E^k + (\mathcal{L}_{Y_f}u)E^k = \big(-ik\alpha_E(Y_f)u + \mathcal{L}_{Y_f}u\big)E^k.$$

Consequently, the Schwartz kernel of $[P_k(f), Q_k(u)]$ is equal to

$$F_k\Big((g - \alpha_E(Y_f))u\Big) + \frac{1}{ik}F_k(\mathcal{L}_{Y_f}u);$$

in other words, $[P_k(f), Q_k(u)] = Q_k\big((g - \alpha_E(Y_f))u\big) + \big(1/(ik)\big)Q_k(\mathcal{L}_{Y_f}u)$. $\qquad\square$

In order to prove Proposition 8.5.3, we will investigate the two terms in the right-hand side of the equality obtained in this lemma. The following result will help us dealing with the first term.

Lemma 8.5.5. *Let* K *be a compact subset of* U. *Then there exists* $C > 0$ *such that for every* $f \in \mathcal{C}^2(M, \mathbb{R})$,

$$|g - \alpha_E(Y_f)| \le C\|f\|_2\varphi_E$$

on K, *with* $g(x, y) = f(x) - f(y)$ *and* $Y_f = (X_f, X_f)$ *as above.*

Proof. Assume first that $K \subset V^2$ where V is a trivialisation open set for M such that $V^2 \subset U$. Introduce some coordinates $x_1, \ldots, x_{2n}$ on V. By Taylor's formula and (8.7), there exist some functions $g_i \in \mathcal{C}^1(V, \mathbb{R})$, $1 \le i \le 2n$, such that for $x, y \in V$

$$g(x, y) = \sum_{i=1}^{2n} g_i(y)(y_i - x_i) + O(\varphi_E)\|f\|_2, \qquad (8.8)$$

and the $O(\varphi_E)$ is uniform on K. Now, write

$$\alpha_E(x,y) = \sum_{j=1}^{2n} \left(\mu_j(x,y)\,\mathrm{d}x_j + \nu_j(x,y)\,\mathrm{d}y_j \right)$$

for some functions $\mu_j, \nu_j \in \mathcal{C}^\infty(V^2)$. Since α_E vanishes along Δ_M, so does μ_j. Therefore, by Taylor's formula, there exist some functions $\mu_{ji} \in \mathcal{C}^\infty(V)$, $1 \le i \le 2n$, such that

$$\mu_j(x,y) = \sum_{i=1}^{2n} \mu_{ji}(y)(y_i - x_i) + O(\varphi_E).$$

Similarly, there exist some functions $\nu_{ji} \in \mathcal{C}^\infty(V)$, $1 \le i \le 2n$, such that

$$\nu_j(x,y) = \sum_{i=1}^{2n} \nu_{ji}(y)(y_i - x_i) + O(\varphi_E).$$

Consequently, we have that

$$\alpha_E(x,y) = \sum_{i=1}^{2n} \left(\sum_{j=1}^{2n} \mu_{ji}(y)\,\mathrm{d}x_j + \nu_{ji}(y)\,\mathrm{d}y_j \right)(y_i - x_i) + O(\varphi_E) \sum_{j=1}^{2n} (\,\mathrm{d}x_j + \,\mathrm{d}y_j).$$

Now, by Taylor's formula, $\mathrm{d}x_j(X_f)(x) = \mathrm{d}x_j(X_f)(y) + O(\varphi_E^{1/2})\|f\|_2$. Thus, the previous formula implies that

$$\alpha_E(Y_f)(x,y) = \sum_{i=1}^{2n} \kappa_i(y)(y_i - x_i) + O(\varphi_E)\|f\|_2 \qquad (8.9)$$

for some smooth functions κ_i, and the $O(\varphi_E)$ is uniform on K. Since, by Lemma 8.5.1, the function $g - \alpha_E(Y_f)$ vanishes to second order along Δ_M, it follows from (8.8) and (8.9) that $g_i - \kappa_i = 0$ for every $i \in [\![1, 2n]\!]$. Therefore

$$g - \alpha_E(Y_f) = O(\varphi_E)\|f\|_2$$

uniformly on K.

To handle the general case, we use the same partition of unity argument that we have used at the end of the proof of Lemma 8.5.2.   □

Finally, the following lemma will take care of the second term in the equality displayed in Lemma 8.5.4.

Lemma 8.5.6. *Let $u \in \mathcal{C}^\infty(M^2, \mathbb{R})$ be a function vanishing to order p along Δ_M. Then there exists $C > 0$ such that for any vector field X of M^2 of class $\mathcal{C}^1$ and tangent to Δ_M, we have that*

$$|\mathcal{L}_X u| \le C\|X\|_1 \varphi_E^{p/2}.$$

Proof. We start by proving the lemma for vector fields which are compactly supported in V^2, where V is a trivialisation open set of M, endowed with coordinates
$x_1, \ldots, x_{2n}$. Write

$$du = \sum_{i=1}^{2n} \left(\frac{\partial u}{\partial x_i}\, dx_i + \frac{\partial u}{\partial y_i}\, dy_i \right) = \sum_{i=1}^{2n} \left(\frac{\partial u}{\partial y_i}(dy_i - dx_i) + \left(\frac{\partial u}{\partial x_i} + \frac{\partial u}{\partial y_i} \right) dx_i \right).$$

Since u vanishes to order p along Δ_M and the vector field $\partial_{x_i} + \partial_{y_i}$ is tangent to Δ_M,
the function $\partial u/\partial x_i + \partial u/\partial y_i$ vanishes to order p along Δ_M, so by Taylor's formula,
it is a $O\big(\varphi_E^{p/2}\big)$. Moreover, there exists $C_1 > 0$ such that for any C^1 vector field X
compactly supported in V^2, $|dx_i(X)| \leq C_1\|X\|_0$. Furthermore, $\partial u/\partial y_i$ vanishes to
order $p - 1$ along Δ_M, so it is a $O\big(\varphi_E^{(p-1)/2}\big)$. We claim that there exists $C_2 > 0$
such that for any C^1 vector field X compactly supported in V^2 and tangent to Δ_M,

$$|(dy_i - dx_i)(X)| \leq C_2\|X\|_1 \varphi_E^{1/2}.$$

Indeed, take any such vector field X and write it as

$$X = \sum_{i=1}^{2n} \alpha_i(x, y)\partial_{x_i} + \beta_i(x, y)\partial_{y_i},$$

where $\alpha_i(x, x) = \beta_i(x, x)$ since X is tangent to Δ_M. Now

$$(dy_i - dx_i)(X) = \beta_i(x, y) - \alpha_i(x, y) = \int_0^1 d(\beta_i - \alpha_i)\big((1 - t)(x, x) + t(x, y)\big)v\, dt$$

with $v = (0, y - x)$, by Taylor's formula. This last term is smaller than a constant
not depending on X times $\|X\|_1 \varphi_E^{1/2}$.

Combining all of the above estimates, we obtain the result for vector fields which
are compactly supported in V^2. We prove the general case by using a partition of
unity argument. $\square$

Let us now show how to apply all of the above.

Proof of Proposition 8.5.3. Let K denote the support of u. Since u vanishes to
order p along the diagonal, it follows from Taylor's formula, (8.7) and a partition
of unity argument that $|u|_{K,p}$ is finite. Consequently, the first estimate follows from
Lemma 8.5.2.

To prove the second estimate, recall that it follows from Lemma 8.5.4 that

$$[P_k(f), Q_k(u)] = Q_k\Big(\big(g - \alpha_E(Y_f)\big)u\Big) + \frac{1}{ik}Q_k\big(\mathcal{L}_{Y_f}u\big).$$

It follows from Lemma 8.5.5 that $|g - \alpha_E(Y_f)| \leq C\|f\|_2\varphi_E$ for some constant $C > 0$
not depending on f. Moreover, since u vanishes to order p along Δ_M, u is a $O\big(\varphi_E^{p/2}\big)$.
Thus, $\big(g - \alpha_E(Y_f)\big)u = O\big(\varphi_E^{(p+2)/2}\big)$, and by Lemma 8.5.2,

$$\left\| Q_k\big((g - \alpha_E(Y_f))u\big) \right\| = O(k^{-p/2-1})\|f\|_2.$$

Similarly, it follows from Lemma 8.5.6 that $|\mathcal{L}_{Y_f} u| \le C'\|f\|_2 \varphi_E^{p/2}$ for some $C' > 0$ not depending on f. Therefore, Lemma 8.5.2 yields

$$\|Q_k(\mathcal{L}_{Y_f} u)\| = O(k^{-p/2})\|f\|_2,$$

and the result follows. $\qquad\qquad\qquad\qquad\qquad\qquad\qquad\qquad\qquad\qquad\qquad\square$

We are now ready to prove Theorem 8.3.1. Write as in Theorem 7.2.1

$$\Pi_k(x,y) = \left(\frac{k}{2\pi}\right)^n E^k(x,y)u(x,y,k) + R_k(x,y),$$

and let $u \sim \sum_{\ell \le 0} k^{-\ell} u_\ell$ be the asymptotic expansion of $u(\,\cdot\,,\cdot\,,k)$. Choose a compactly supported function $\chi \in \mathcal{C}^\infty(M^2, \mathbb{R})$ such that $\mathrm{supp}(\chi) \subset U$ and equal to one near Δ_M. Fixing $m \in \mathbb{N}$, we write

$$\Pi_k = \sum_{\ell=0}^m k^{-\ell} Q_k(\chi u_\ell) + \sum_{\ell=0}^m k^{-\ell} Q_k\big((1-\chi)u_\ell\big) + Q_k\left(u - \sum_{\ell=0}^m k^{-\ell} u_\ell\right) + R_k,$$

where R_k is the operator with Schwartz kernel $R_k(\cdot,\cdot)$. We only need to estimate the commutator of each of these terms with $P_k(f)$. Since χu_ℓ is compactly supported in U, it follows from Proposition 8.5.3 that $[P_k(f), Q_k(\chi u_\ell)] = O(k^{-1})\|f\|_2$, so

$$\left[P_k(f), \sum_{\ell=0}^m k^{-\ell} Q_k(\chi u_\ell) \right] = O(k^{-1})\|f\|_2.$$

For the second term, we use the following fact. Let V be a neighbourhood of Δ_M, and let $r = \sup_{M^2 \setminus V} \|E\| < 1$; then for any $v \in \mathcal{C}^0(M^2)$ vanishing in V, we have that

$$\|F_k(v)\| \le C k^n r^k \|v\|_0$$

for some $C > 0$ not depending on v. Therefore this Schwartz kernel is a $O(k^{-\infty})\|v\|_0$ uniformly on M^2, and by Proposition 6.4.1, $Q_k(v) = O(k^{-\infty})\|v\|_0$. Since $1 - \chi$ vanishes in a neighbourhood of Δ_M, combining this fact with the equality

$$[P_k(f), Q_k((1-\chi)u_\ell)] = Q_k\big((1-\chi)(g - \alpha_E(Y_f))u_\ell\big) + \frac{1}{ik} Q_k\big(\mathcal{L}_{Y_f}((1-\chi)u_\ell)\big),$$

coming from Lemma 8.5.4, we obtain that

$$\left[P_k(f), \sum_{\ell=0}^m k^{-\ell} Q_k\big((1-\chi)u_\ell\big) \right] = O(k^{-1})\|f\|_2.$$

It only remains to estimate the commutator $[P_k(f), S_k]$ where

$$S_k = Q_k \left(u - \sum_{\ell=0}^{m} k^{-\ell} u_\ell \right) + R_k.$$

The Schwartz kernel $S_k(\,\cdot\,,\cdot\,)$ of S_k is a $O(k^{n-(m+1)})$. We conclude the proof by taking m large enough and using the following lemma.

Lemma 8.5.7. *There exists $C > 0$ such that for every $f \in \mathcal{C}^2(M, \mathbb{R})$,*

$$\|[P_k(f), S_k]\| \leq C k^{n-(m+1)} \|f\|_2.$$

Proof. By computing $\widetilde{\nabla}^k \big(F_k(u - \sum_{\ell=0}^{m} k^{-\ell} u_\ell) \big)$, we obtain that for every vector field X on M^2 of class $\mathcal{C}^0$, there exists $C_X > 0$ such that $\|\widetilde{\nabla}_X^k S_k\| \leq C_X k^{n-m}$. This implies that there exists $C > 0$ such that for every vector field X on M^2 of class $\mathcal{C}^0$, the inequality $\|\widetilde{\nabla}_X^k S_k\| \leq C k^{n-m} \|X\|_0$ holds. Indeed, let $(\eta_i)_{1 \leq i \leq q}$ be a partition of unity subordinate to an open cover $(U_i)_{1 \leq i \leq q}$ of M^2 by trivialisation open sets for TM^2, with a local basis $(Y_{ij})_{1 \leq j \leq 4n}$, and write

$$X = \sum_{i=1}^{q} \eta_i X = \sum_{i=1}^{q} \sum_{j=1}^{4n} \lambda_{ij} Y_{ij},$$

where λ_{ij} is a continuous function, which satisfies $\|\lambda_{ij}\|_0 \leq C' \|X\|_0$ for some $C' > 0$. Consequently,

$$\|\widetilde{\nabla}_X^k S_k\| = \left\| \sum_{i=1}^{q} \sum_{j=1}^{4n} \lambda_{ij} \widetilde{\nabla}_{Y_{ij}}^k S_k \right\| \leq C' (\max_{i,j} C_{Y_{ij}}) \|X\|_0.$$

To finish the proof, we obtain as in the proof of Lemma 8.5.4 that the Schwartz kernel of $[P_k(f), S_k]$ is equal to

$$\left(f \boxtimes \mathrm{id} - \mathrm{id} \boxtimes f + \frac{1}{ik} \widetilde{\nabla}_{(X_f, X_f)}^k \right) S_k.$$

By the above estimate, $\|\widetilde{\nabla}_{(X_f, X_f)}^k S_k\| \leq C k^{n-m} \|f\|_1$, and the result follows. $\qquad\square$

Chapter 9
Coherent States and Norm Correspondence

Finally, we prove the lower bound for the operator norm of a Berezin–Toeplitz operator. In order to do so, we use the so-called coherent states.

9.1 Coherent Vectors

Let $P \subset L$ be the set of elements $u \in L$ such that $\|u\| = 1$, and denote by $\pi \colon P \to M$ the natural projection.

Lemma 9.1.1. *Fix $u \in P$. For every $k \geq 1$, there exists a unique vector ξ_k^u in $\mathcal{H}_k$ such that*
$$\forall \phi \in \mathcal{H}_k, \quad \phi\big(\pi(u)\big) = \langle \phi, \xi_k^u \rangle_k u^k.$$

Definition 9.1.2. The vector $\xi_k^u \in \mathcal{H}_k$ is called the *coherent vector* at u.

Proof of Lemma 9.1.1. Consider the linear form F_k defined on $\mathcal{H}_k$ by
$$\forall \phi \in \mathcal{H}_k, \quad F_k(\phi) = h_k\big(\phi\big(\pi(u)\big), u^k\big).$$

Since $\mathcal{H}_k$ is finite-dimensional, F_k is continuous, so the Riesz representation theorem implies that there exists a unique vector $\xi_k^u \in \mathcal{H}_k$ such that $F_k(\phi) = \langle \phi, \xi_k^u \rangle_k$ for all ϕ in $\mathcal{H}_k$. But since u^k is an orthonormal basis of $L_{\pi(u)}^k$, we have $\phi\big(\pi(u)\big) = F_k(\phi)u^k$. $\square$

Lemma 9.1.3. *Let T_k be an operator $\mathcal{C}^\infty(M, L^k) \to \mathcal{C}^\infty(M, L^k)$ with kernel $T_k(\cdot, \cdot)$ and such that $\Pi_k T_k \Pi_k = T_k$. Then*

(1) $\forall x \in M,\ (T_k \xi_k^u)(x) = T_k\big(x, \pi(u)\big) \cdot u^k$,
(2) $\langle T_k \xi_k^u, \xi_k^v \rangle_k = \bar{v}^k \cdot T_k\big(\pi(v), \pi(u)\big) \cdot u^k$,

where we recall that the dot stands for contraction with respect to h_k.

© Springer International Publishing AG, part of Springer Nature 2018
Y. Le Floch, *A Brief Introduction to Berezin–Toeplitz Operators on Compact Kähler Manifolds*, CRM Short Courses,
https://doi.org/10.1007/978-3-319-94682-5_9

Proof. Let $(\phi_i)_{1 \leq i \leq d_k}$ be an orthonormal basis of $\mathcal{H}_k$. By proposition 6.3.3, we can write the Schwartz kernel of the restriction of T_k to $\mathcal{H}_k$ as

$$\forall x, y \in M, \quad T_k(x, y) = \sum_{i,j=1}^{d_k} \langle T_k \phi_i, \phi_j \rangle_k \phi_j(x) \otimes \overline{\phi_i(y)}.$$

Therefore, for $x \in M$ we have that

$$T_k\big(x, \pi(u)\big) \cdot u^k = \sum_{i,j=1}^{d_k} \langle T_k \phi_i, \phi_j \rangle_k h_k\Big(u^k, \phi_i(\pi(u))\Big) \phi_j(x),$$

which we can rewrite, because $h_k\big(u^k, \phi_i(\pi(u))\big) = \langle \xi_k^u, \phi_i \rangle_k$, as

$$T_k\big(x, \pi(u)\big) \cdot u^k = \sum_{j=1}^{d_k} \langle T_k \left(\sum_{i=1}^{d_k} \langle \xi_k^u, \phi_i \rangle_k \phi_i \right), \phi_j \rangle_k \phi_j(x),$$

which yields that

$$T_k\big(x, \pi(u)\big) \cdot u^k = \sum_{j=1}^{d_k} \langle T_k \xi_k^u, \phi_j \rangle_k \phi_j(x) = (T_k \xi_k^u)(x).$$

This corresponds to the first claim. For the second claim, we use the first one to write for x in M that $h_k\big((T_k \xi_k^u)(x), \xi_k^v(x)\big) = h_k\big(T_k(x, \pi(u)) \cdot u^k, \xi_k^v(x)\big)$. Integrating this equality leads to

$$\langle T_k \xi_k^u, \xi_k^v \rangle_{\mathcal{H}_k} = \langle T_k(\,\cdot\,, \pi(u)), \xi_k^v \rangle_k,$$

but the right-hand side of this equation is equal to $h_k\big(T_k(\pi(v), \pi(u)) \cdot u^k, \xi_k^v(x)\big)$ by definition of ξ_k^v, and this term is in turn equal to $\bar{v}^k \cdot T_k(\pi(v), \pi(u)) \cdot u^k$. $\square$

By taking $T_k = \Pi_k$ in this proposition, we immediately get the following properties.

Corollary 9.1.4. *For every* $u, v \in P$,

(1) *for every* x *in* M, $\xi_k^u(x) = \Pi_k\big(x, \pi(u)\big) \cdot u^k$,
(2) $\langle \xi_k^u, \xi_k^v \rangle_k = \bar{v}^k \cdot \Pi_k\big(\pi(v), \pi(u)\big) \cdot u^k$, *so* $\Pi_k\big(\pi(v), \pi(u)\big) = \langle \xi_k^u, \xi_k^v \rangle_k v^k \otimes \bar{u}^k$,
(3) $\|\xi_k^u\|_k^2 = \Pi_k\big(\pi(u), \pi(u)\big)$.

9.2 Operator Norm of a Berezin–Toeplitz Operator

In this section, we prove Theorem 5.2.1. By the above corollary and Theorem 7.2.1, we have that for every $u \in P$,

$$\|\xi_k^u\|_k^2 \sim \left(\frac{k}{2\pi}\right)^n$$

when k goes to infinity, the estimate being uniform in u. In particular, there exists $k_0 \geq 1$ such that for every $u \in P$, $\xi_k^u \neq 0$ whenever $k \geq k_0$. For $k \geq k_0$, we set $\xi_k^{u,\mathrm{norm}} = \xi_k^u/\|\xi_k^u\|_k$. Observe also that this means that the class of ξ_k^u in the projective space $\mathbb{P}(\mathcal{H}_k)$ is well-defined. In fact, this class only depends on $\pi(u)$ (because for $\lambda \in \mathbb{C}$, $\xi_k^{\lambda u} = \lambda^k \xi_k^u$) and is called the *coherent state* at $x = \pi(u)$.

Proposition 9.2.1. *There exists $C > 0$ such that for every $x \in M$, for every $u \in P$ such that $x = \pi(u)$ and for every $f \in \mathcal{C}^2(M,\mathbb{R})$ having x as a critical point,*

$$\|T_k(f)\xi_k^{u,\mathrm{norm}} - f(x)\xi_k^{u,\mathrm{norm}}\|_k \leq Ck^{-1}\|f\|_2$$

for every $k \geq k_0$.

Proof. Let $(U_i)_{1 \leq i \leq m}$ be an open cover of M by trivialisation open sets, and let $(V_i)_{1 \leq i \leq m}$ be a refinement of $(U_i)_{1 \leq i \leq m}$ such that $\overline{V_i} \subset U_i$ is compact. Then it is enough to show that for every $i \in [\![1,m]\!]$, there exists $C_i > 0$ such that for every $x \in V_i$, for every $u \in P$ such that $x = \pi(u)$ and for every $f \in \mathcal{C}^2(M,\mathbb{R})$ having x as a critical point,

$$\|T_k(f)\xi_k^{u,\mathrm{norm}} - f(x)\xi_k^{u,\mathrm{norm}}\|_k \leq Ck^{-1}\|f\|_2$$

for every $k \geq k_0$. Indeed it will then suffice to take $C = \max_{1 \leq i \leq m} C_i$. So let us choose $i \in [\![1,d]\!]$ and let us take $x \in V_i$, and set $\lambda = f(x)$. Then

$$\|(f - \lambda)\xi_k^{u,\mathrm{norm}}\|_k^2 = \int_{V_i} |f(y) - \lambda|^2 \, \|\xi_k^{u,\mathrm{norm}}(y)\|^2 \mu(y)$$
$$+ \int_{M \setminus V_i} |f(y) - \lambda|^2 \|\xi_k^{u,\mathrm{norm}}(y)\|^2 \mu(y).$$

We will estimate both integrals. Let us introduce some coordinates $y_1, \ldots, y_{2n}$ on U_i such that $x = (0, \ldots, 0)$, and set $q(y) = \sum_{j=1}^{2n} y_j^2$. By Taylor's formula, there exists a constant $\alpha > 0$, not depending on f, such that $|f(y) - \lambda| \leq \alpha\|f\|_2 q(y)$ for every $y \in V_i$. Therefore,

$$\int_{V_i} |f(y) - \lambda|^2 \|\xi_k^{u,\mathrm{norm}}(y)\|^2 \mu(y) \leq \alpha^2 \|f\|_2^2 \int_{V_i} \|\xi_k^{u,\mathrm{norm}}(y)\|^2 q(y)^2 \, \mu(y).$$

In order to estimate this integral, we write:

$$\|\xi_k^{u,\mathrm{norm}}(y)\| = \frac{\|\xi_k^u(y)\|}{\|\xi_k^u\|_k} = \frac{\|\Pi_k(y,x) \cdot u^k\|}{\|\xi_k^u\|_k}.$$

We claim that $\|\Pi_k(y,x) \cdot u^k\| = \|\Pi_k(y,x)\|$. This is easily proved by fixing y, taking $v \in L_y$ with unit norm, and writing $\Pi_k(y,x)$ in the orthonormal basis $v^k \otimes \bar{u}^k$ of $L_y^k \otimes \bar{L}_x^k$. But it follows from (8.7) that there exists $\beta > 0$ such that for every $y \in V_i$,

$\|E(y,x)\| \leq \exp\bigl(-\beta q(y)\bigr)$. Therefore, using Theorem 7.2.1 and remembering that $\|\xi_k^u\|_k^2 \sim \bigl(k/(2\pi)\bigr)^n$, we obtain that there exists $\gamma > 0$ independent of f, x and u such that

$$\forall y \in V_i, \quad \|\xi_k^{u,\mathrm{norm}}(y)\|^2 \leq \gamma k^n \exp\bigl(-2\beta k q(y)\bigr).$$

Now, on U_i we can write $\mu = g\,dy_1 \wedge \cdots \wedge dy_{2n}$ for some smooth function g. So, if $\delta = \max_{\overline{V_i}}|g|$, we have that

$$\int_{V_i} \|\xi_k^{u,\mathrm{norm}}(y)\|^2 q(y)^2 \mu(y) \leq \gamma \delta k^n \int_{\mathbb{R}^{2n}} \exp\bigl(-2\beta k q(y)\bigr) q(y)^2 \, dy.$$

By performing the change of variable $w = \sqrt{k}\, y$, we finally obtain that

$$\int_{V_i} \|\xi_k^{u,\mathrm{norm}}(y)\|^2 q(y)^2 \mu(y) \leq \varepsilon k^{-2}$$

for some $\varepsilon > 0$, not depending on f, x, u. Consequently,

$$\int_{V_i} |f(y) - \lambda|^2 \|\xi_k^{u,\mathrm{norm}}(y)\|^2 \mu(y) \leq \alpha^2 \varepsilon \|f\|_2^2 k^{-2}.$$

It remains to estimate the integral on $M \setminus V_i$. Since for every $y \in M$, we have that $|f(y) - \lambda| \leq 2\|f\|_0 \leq 2\|f\|_2$, we immediately obtain that

$$\int_{M\setminus V_i} |f(y) - \lambda|^2 \|\xi_k^{u,\mathrm{norm}}(y)\|^2 \mu(y) \leq 4\|f\|_2^2 \int_{M\setminus V_i} \|\xi_k^{u,\mathrm{norm}}(y)\|^2 \, \mu(y).$$

We claim that this last integral is a $O\bigl(k^{-2}\bigr)$. This comes again from the fact that $\|\xi_k^{u,\mathrm{norm}}(y)\| = \|\Pi_k(y,x)\|/\|\xi_k^u\|_k$, since there exists $r < 1$ such that $\|E(y,x)\| \leq r$ whenever y belongs to $M \setminus V_i$. So we finally get that

$$\|(f - \lambda)\xi_k^{u,\mathrm{norm}}\|_k \leq C_i \|f\|_2 k^{-1}$$

for some $C_i > 0$ independent of f, x, u. Since the operator norm of Π_k is smaller than one, this yields

$$\|(T_k(f) - \lambda)\xi_k^{u,\mathrm{norm}}\|_k = \|\Pi_k(f - \lambda)\xi_k^{u,\mathrm{norm}}\|_k \leq C_i \|f\|_2 k^{-1},$$

which concludes the proof. $\square$

To prove Theorem 5.2.1, we assume that the maximum of $|f|$ is $f(x_0)$ for some $x_0 \in M$ (otherwise, we work with $-f$), and we apply the previous result to x_0 and $u \in L_{x_0}$. This gives

$$\|T_k(f)\xi_k^{u,\mathrm{norm}} - \|f\|\xi_k^{u,\mathrm{norm}}\| \leq Ck^{-1}\|f\|_2.$$

This implies that the distance between $\|f\|$ and the spectrum of $T_k(f)$ satisfies $\mathrm{dist}\bigl(\|f\|, \mathrm{Sp}(T_k(f))\bigr) \leq Ck^{-1}\|f\|_2$. Indeed, it is an easy consequence of the spectral theorem that if A is a bounded self-adjoint operator acting on a Hilbert space, then

$$\|(A - \lambda)^{-1}\| \leq \frac{1}{\mathrm{dist}\left(\lambda, \mathrm{Sp}(A)\right)}$$

for every $\lambda \notin \mathrm{Sp}(A)$. So there exists $\lambda \in \mathrm{Sp}\left(T_k(f)\right)$ such that $\lambda \geq \|f\| - Ck^{-1}\|f\|_2$. Therefore, we have that

$$\|T_k(f)\| = \max_{\mu \in \mathrm{Sp}(T_k(f))} |\mu| \geq \|f\| - Ck^{-1}\|f\|_2.$$

9.3 Positive Operator-Valued Measures

Let us show how the coherent states that we have introduced can be used to describe Berezin–Toeplitz operators in terms of integrals against a positive operator-valued measure. Firstly, let us recall what this term means. Let $\mathcal{H}$ be a complex Hilbert space, and let $\mathcal{S}(\mathcal{H})$ be the space of bounded self-adjoint operators on $\mathcal{H}$. Let X be a set endowed with a σ-algebra $\mathcal{C}$.

Definition 9.3.1. A *positive operator-valued measure* on X with values in $\mathcal{S}(\mathcal{H})$ is a map $G \colon \mathcal{C} \to \mathcal{S}(\mathcal{H})$ which satisfies the following properties:

(1) for every $A \in \mathcal{C}$, $G(A)$ is a positive operator, i.e. $\langle A\xi, \xi \rangle \geq 0$ for every $\xi \in \mathcal{H}$,
(2) $G(\varnothing) = 0$ and $G(X) = \mathrm{Id}$,
(3) G is σ-additive: for any sequence $(A_j)_{j \geq 1}$ of disjoint elements of $\mathcal{C}$, $G\left(\bigcup_{j \geq 1} A_j\right) = \sum_{j \geq 1} G(A_j)$.

Such a positive operator-valued measure defines a probability measure μ_ξ on X for every $\xi \in \mathcal{H}$, by the formula $\mu_\xi(A) = \langle G(A)\xi, \xi \rangle$ for $A \in \mathcal{C}$. Given a bounded measurable function $f : X \to \mathbb{R}$, we define an operator $\int_X f dG \in \mathcal{S}(\mathcal{H})$ characterised by the following property:

$$\forall \xi \in \mathcal{H}, \quad \left\langle \left(\int_X f\, dG\right)\xi, \xi \right\rangle = \int_X f\, d\mu_\xi.$$

Coming back to the context of Berezin–Toeplitz operators, we consider $X = M$ with the σ-algebra generated by its Borel sets, and $\mathcal{H} = \mathcal{H}_k = H^0(M, L^k)$. As before, for $x \in M$ and $u \in L_x$ with unit norm, let ξ_k^u be the coherent vector at u. Recall that there exists $k_0 \geq 1$ such that $\xi_k^u \neq 0$ whenever $k \geq k_0$. We claim that the function

$$\rho_k \colon M \to \mathbb{R}, \quad x \mapsto \|\xi_k^u\|_k^2$$

is well-defined, i.e. only depends on x. Indeed, if v is another unit vector in L_x, then $v = \lambda u$ for some $\lambda \in \mathbb{S}^1$. But then we have that $\xi_k^v = \lambda^k \xi_k^u$, so $\|\xi_k^v\|_k^2 = \|\xi_k^u\|_k^2$. For $k \geq k_0$, ρ_k is a positive function. Furthermore, the projection

$$P_k^x : \mathcal{H}_k \to \mathcal{H}_k, \quad \phi \mapsto \frac{\langle \phi, \xi_k^u \rangle_k}{\|\xi_k^u\|_k^2} \xi_k^u$$

is also only dependent on x.

Lemma 9.3.2. *For $k \geq k_0$, the map G_k such that $\mathrm{d}G_k = \rho_k(x)P_k^x \mu$ defines a positive operator-valued measure on M.*

Proof. The positivity and σ-additivity are immediate from the form of G_k. Let us prove the fact that $G_k(M) = \mathrm{Id}$. Let $\phi \in \mathcal{H}_k$ and $y \in M$; we have that

$$(G_k(M)\phi)(y) = \int_M \rho_k(x)(P_k^x \phi)(y)\mu(x).$$

Recall that $\xi_k^u(y) = \Pi_k(y,x) \cdot u^k$. Thus,

$$\rho_k(x)(P_k^x \phi)(y) = \langle \phi, \xi_k^u \rangle_k \xi_k^u(y) = \Pi_k(y,x) \cdot \left(\langle \phi, \xi_k^u \rangle_k u^k \right).$$

But ξ_k^u satisfies the reproducing property (7.2), hence $\langle \phi, \xi_k^u \rangle_k u^k = \phi(x)$. So finally

$$(G_k(M)\phi)(y) = \int_M \Pi_k(y,x) \cdot \phi(x)\mu(x) = (\Pi_k \phi)(y) = \phi(y). \qquad \square$$

Proposition 9.3.3. *Let $k \geq k_0$. For any $f \in \mathcal{C}^\infty(M, \mathbb{R})$, $T_k(f) = \int_M f \, \mathrm{d}G_k$.*

Proof. Let $S_k(f) = \int_M f \, \mathrm{d}G_k$, and let $\phi \in \mathcal{H}_k$. Then by definition,

$$\langle S_k(f)\phi, \phi \rangle_k = \int_M f(x)\langle P_k^x \phi, \phi \rangle_k \rho_k(x)\mu(x).$$

We claim that for every $x \in M$, $\langle P_k^x \phi, \phi \rangle_k \rho_k(x) = h_k\big(\phi(x), \phi(x)\big)$. Indeed, on the one hand, since ξ_k^u satisfies the reproducing property (7.2), we have that $\phi(x) = \langle \phi, \xi_k^u \rangle_k u^k$. Therefore

$$h_k\big(\phi(x), \phi(x)\big) = |\langle \phi, \xi_k^u \rangle_k|^2 h_k(u^k, u^k) = |\langle \phi, \xi_k^u \rangle_k|^2.$$

But on the other hand, we have that

$$\langle P_k^x \phi, \phi \rangle_k = \frac{|\langle \phi, \xi_k^u \rangle_k|^2}{\|\xi_k^u\|_k^2} = \frac{|\langle \phi, \xi_k^u \rangle_k|^2}{\rho_k(x)},$$

which proves the claim. Consequently,

$$\langle S_k(f)\phi, \phi \rangle_k = \int_M h_k\big((f(x)\phi(x), \phi(x)\big)\mu(x) = \langle T_k(f)\phi, \phi \rangle_k,$$

which proves the result. $\qquad \square$

9.4 Projective Embeddings

The coherent states construction gives a way to embed M into a complex projective space. Remember that given a unit vector $u \in L$, the coherent state $\xi_k^u \in \mathcal{H}_k$ at u is the holomorphic section of $L^k \to M$ given by

$$\xi_k^u(y) = \Pi_k\big(y, \pi(u)\big) \cdot u^k,$$

and that there exists $k_0 \geq 1$ such that for every $k \geq k_0$ and for every unit vector $u \in L$, $\xi_k^u \neq 0$. Hence for $k \geq k_0$ (from now on, we will assume that it is the case), the class $[\xi_k^u]$ of ξ_k^u in $\mathbb{P}(\mathcal{H}_k)$ is well-defined, and we saw that this class only depends on $\pi(u)$ where π is the projection from L to M. Thus we obtain a map

$$\Phi_{\mathrm{coh}} \colon M \to \mathbb{P}(\mathcal{H}_k), \qquad x \mapsto [\xi_k^u], \quad u \in \pi^{-1}(x).$$

Since $\Pi(\cdot, \cdot)$ is anti-holomorphic on the right variable, this map is anti-holomorphic. To get a holomorphic map, we consider

$$\Phi_{\mathrm{hol}} \colon M \to \mathbb{P}(\mathcal{H}_k^*), \qquad x \mapsto [\langle \cdot, \xi_k^u \rangle_k], \quad u \in \pi^{-1}(x).$$

By Lemma 9.1.1, we have the alternative expression $\Phi_{\mathrm{hol}}(x) = [\alpha_u]$ for any $u \in \pi^{-1}(x)$ with norm one, where $\alpha_u(\phi) = \phi(x) \cdot \bar{u}^k$ for every $\phi \in \mathcal{H}_k$.

In order to identify $\mathbb{P}(\mathcal{H}_k)$ with $\mathbb{C}\mathbb{P}^{d_k}$, let us choose an orthonormal basis $(\varphi_j)_{0 \leq j \leq d_k}$ of $\mathcal{H}_k$, $d_k = \dim(\mathcal{H}_k) - 1$, and let us write for any unit vector $u \in L$

$$\xi_k^u = \sum_{j=0}^{d_k} \lambda_j(u) \varphi_j$$

for some complex numbers $\lambda_0(u), \ldots, \lambda_{d_k}(u)$. Then, using homogeneous coordinates,

$$\Phi_{\mathrm{coh}}(x) = [\lambda_0(u) : \cdots : \lambda_{d_k}(u)], \quad \Phi_{\mathrm{hol}}(x) = \big[\overline{\lambda_0(u)} : \cdots : \overline{\lambda_{d_k}(u)}\big].$$

The latter is obtained by decomposing $\langle \cdot, \xi_k^u \rangle$ in the dual basis $(\varphi_j^*)_{0 \leq j \leq d_k}$.

Proposition 9.4.1. *The maps Φ_{coh} and Φ_{hol} are embeddings for k large enough.*

Proof. Since L^k is very ample for k large enough because L is positive, this follows from the fact that Φ_{hol} is the embedding considered in Kodaira's embedding theorem [24, Section 5.3]. Indeed, for $j \in [\![0, d_k]\!]$ and $x \in M$, we have that for any unit vector $u \in \pi^{-1}(x)$:

$$\varphi_j(x) = \langle \varphi_j, \xi_k^u \rangle_k u^k = \overline{\lambda_j(u)} u^k. \qquad \square$$

As before, let $\rho_k \colon M \to \mathbb{R}$ be the function sending $x \in M$ to $\|\xi_k^u\|_k^2$ for any $u \in L_x$ with norm one. This function is often called *Rawnsley's function*, since it was introduced in [40] (see also [39]); however, the reader may encounter this terminology for a slightly different function, since many authors work with elements $u \neq 0 \in L$ instead of unit vectors.

Proposition 9.4.2. *The pullback of the Fubini–Study form by Φ_{hol} is given by*

$$\Phi_{\mathrm{hol}}^* \omega_{\mathrm{FS}} = k\omega + i\partial\bar{\partial}\log\rho_k.$$

Proof. As in Example 2.5.9, introduce, for $j \in [\![1, d_k]\!]$, the open subset

$$U_j = \{[z_0 : \cdots : z_{d_k}] \in \mathbb{CP}^{d_k} \mid z_j \neq 0\}$$

of $\mathbb{CP}^{d_k}$. Then on U_j,

$$\omega_{\mathrm{FS}} = i\partial\bar{\partial}\log\left(\sum_{m=0}^{d_k}\left|\frac{z_m}{z_j}\right|^2\right).$$

Therefore, we have that, on U_j:

$$\Phi_{\mathrm{hol}}^*\omega_{\mathrm{FS}} = i\partial\bar{\partial}\log\left(\sum_{m=0}^{d_k}\left|\frac{\lambda_m}{\lambda_j}\right|^2\right) = i\partial\bar{\partial}\log\rho_k - i\partial\bar{\partial}\log|\lambda_j|^2. \qquad (9.1)$$

Now, let u_j be a local section of L over U_j such that $u_j(x)$ is a unit vector of L_x for every $x \in U_j$. Then $\varphi_j(x) = \lambda_j\big(u_j(x)\big)u_j(x)^k$ is a local non-vanishing holomorphic section of L, thus, remembering the proof of Proposition (3.5.4), we get that

$$\nabla^k\varphi_j = \beta_j \otimes \varphi_j, \quad \beta_j = \partial\log H_j$$

on U_j, with $H_j = h_k(\varphi_j, \varphi_j) = |\lambda_j(u_j)|^2$. Therefore

$$-ik\omega = \mathrm{curv}(\nabla^k) = \bar{\partial}\partial\log H_j = \bar{\partial}\partial\log|\lambda_j(u_j)|^2$$

on U_j, which, in view of (9.1), yields the result.   □

Thus $\Phi_{\mathrm{hol}}^*\omega_{\mathrm{FS}} = k\omega$ whenever ρ_k is constant. In this case, applying Proposition 9.3.3 to $f = 1$, we get that

$$\dim\mathcal{H}_k = \int_M \rho_k\mu(x) = \mathrm{vol}(M)\rho_k,$$

therefore $\rho_k = \dim\mathcal{H}_k / \mathrm{vol}(M)$.

Example 9.4.3 (The complex projective line). Let us come back to Example 7.2.5. On $U_0 = \{[z_0 : z_1] \mid z_0 \neq 0\}$, we have the following expression for the kernel of Π_k:

$$\Pi_k(z, w) = \frac{k+1}{2\pi}(1 + z\bar{w})^k\, t_0^k(z) \otimes \bar{t}_0^k(w).$$

Considering the unit vector

$$u(z) = \frac{1}{h\big(t_0(z), t_0(z)\big)^{1/2}} t_0(z) = \sqrt{1 + |z|^2}\, t_0(z),$$

we get that the coherent state at $u(z)$ has value at w

$$\xi_k^{u(z)}(w) = \frac{k+1}{2\pi} \big(1 + |z|^2\big)^{k/2} (1 + \bar{z}w)^k h\big(t_0(z), t_0(z)\big)^k t_0^k(w)$$

$$= \frac{k+1}{2\pi} \frac{(1 + \bar{z}w)^k}{(1 + |z|^2)^{k/2}} t_0^k(w).$$

Exercise 9.4.4. Check that $\rho_k(z) = \|\xi_k^{u(z)}\|_k^2 = (k+1)/(2\pi)$.

To understand the coherent states embedding, we expand this coherent state to get a linear combination of the $e_\ell(w) = \sqrt{(k+1)\binom{k}{\ell}/(2\pi)}\, w^{k-\ell} t_0^k(w)$, $0 \le \ell \le k$:

$$\xi_k^{u(z)}(w) = \sqrt{\frac{(k+1)}{2\pi(1 + |z|^2)^k}} \sum_{\ell=0}^k \sqrt{\binom{k}{\ell}}\, \bar{z}^\ell e_\ell(w).$$

This means that

$$\Phi_{\mathrm{coh}}(z) = \left[1 : \cdots : \sqrt{\binom{k}{\ell}}\, \bar{z}^\ell : \cdots : \bar{z}^k \right]$$

and finally

$$\Phi_{\mathrm{hol}}(z) = \left[1 : \cdots : \sqrt{\binom{k}{\ell}}\, z^\ell : \cdots : z^k \right]$$

is the Veronese embedding of $\mathbb{CP}^1$ into $\mathbb{CP}^k$.

Appendix A
The Circle Bundle Point of View

The goal of this appendix is to compare the line bundle version of geometric quanti-sation and Berezin–Toeplitz operators with the circle bundle version of this theory. To this effect, we begin by recalling some useful facts about $\mathbb{T}$-principal bundles with connections. Then, we discuss the Hardy space and the Szegő projector of a strictly pseudoconvex domain. Finally, we explain how this enters the picture of geometric quantisation. For this appendix, we assume from the reader a basic knowledge of Lie groups and their representations.

A.1 $\mathbb{T}$-Principal Bundles and Connections

Let G be a Lie group and let X be a manifold.

Definition A.1.1. A *G-principal bundle* over X (or principal bundle over X with structure group G) is the data of a manifold P (the total space) and a smooth projection $\pi : P \to X$ together with an action of G on P such that

(1) G acts freely and transitively on P on the right: $(p, g) \in P \times G \mapsto pg \in P$,
(2) X is the quotient of P by the equivalence relation induced by this action, and π is the canonical projection,
(3) P is locally trivial in the sense that each point $x \in X$ has a neighbourhood U such that there exists a diffeomorphism

$$\varphi : \pi^{-1}(U) \to U \times G$$

of the form $\varphi(p) = \big(\pi(p), \psi(p)\big)$, where the map $\psi : \pi^{-1}(U) \to G$ is such that $\psi(pg) = \psi(p)g$ for every $p \in \pi^{-1}(U)$ and $g \in G$.

Let $P \to X$ be a principal bundle with structure group G, and let $\phi : G \to \mathrm{GL}(V)$ be a representation of G on some vector space V. There is a free action of G on $P \times V$ on the right:

© Springer International Publishing AG, part of Springer Nature 2018   125
Y. Le Floch, *A Brief Introduction to Berezin–Toeplitz Operators
on Compact Kähler Manifolds*, CRM Short Courses,
https://doi.org/10.1007/978-3-319-94682-5

$$(p, v, g) \in P \times V \times G \mapsto (p, g)v := \big(pg, \phi(g^{-1})v\big) \in P \times V.$$

This action induces an equivalence relation on $P \times V$; by taking the quotient, we obtain a vector bundle $(P \times V)/G \to P/G = X$ whose fibres $(G \times V)/G$ are isomorphic to V.

Definition A.1.2. We denote by $P \times_\phi V \to X$ the vector bundle $(P \times V)/G \to X$, and we call it the vector bundle associated with the G-principal bundle $P \to X$ and the representation ϕ.

$\mathbb{T}$-Principal Bundles

Let $P \to X$ be a principal bundle with structure group $\mathbb{T} = \mathbb{R}/2\pi\mathbb{Z}$ and projection π. The action of $\theta \in \mathbb{T}$ will be denoted by

$$(p, \theta) \in P \times \mathbb{T} \mapsto R_\theta(p) \in P.$$

To this action is associated the vector field ∂_θ of P defined as

$$\forall p \in P \quad \partial_\theta(p) = \frac{\mathrm{d}}{\mathrm{d}t}\bigg|_{t=0} R_t(p)$$

whose flow at time t is equal to R_t. The elements of $\ker(d_p\pi) = \mathrm{span}\big(\partial_\theta(p)\big)$ are called the *vertical* tangent vectors.

Definition A.1.3. A *connection* on $P \to X$ is the data of a one-form $\alpha \in \Omega^1(P)$ which is $\mathbb{T}$-invariant ($R_\theta^*\alpha = \alpha$ for every $\theta \in \mathbb{T}$) and satisfies $i_{\partial_\theta}\alpha = 1$.

A connection $\alpha \in \Omega^1(P)$ induces a splitting

$$T_pP = \ker(\alpha_p) \oplus \mathrm{span}\big(\partial_\theta(p)\big) = \ker(\alpha_p) \oplus \ker(d_p\pi).$$

The elements of the hyperplane $\ker(\alpha_p)$ of T_pP are called the *horizontal* tangent vectors. Since α is $\mathbb{T}$-invariant, the distribution $\ker \alpha$ also is, and the data of a connection is equivalent to the data of a $\mathbb{T}$-invariant subbundle E of TP such that $TP = E \oplus \ker(d\pi)$. By construction, the restriction of $d_p\pi$ to the horizontal subspace at p is bijective. Thus, given a vector field Y on X, there exists a unique vector field Y^{hor} on P which is horizontal and satisfies $d\pi(Y^{\mathrm{hor}}) = Y$; it is called the *horizontal lift* of Y.

The connections of the trivial $\mathbb{T}$-principal bundle $X \times \mathbb{T}$ are the one-forms of the type $\beta + \mathrm{d}\theta$, where $\beta \in \Omega^1(X)$ and $\mathrm{d}\theta$ is the usual 1-form of $\mathbb{T}$.

$\mathbb{T}$-Principal Bundles and Hermitian Line Bundles

Let $L \to X$ be a Hermitian complex line bundle, and let $h(\cdot, \cdot)$ denote its Hermitian form. Let us consider the subbundle of L consisting of elements of norm 1:

$$P = \{u \in L \mid h(u, u) = 1\}.$$

One readily checks that P is a $\mathbb{T}$-principal bundle over X, with $\mathbb{T}$-action given by $R_\theta(u) = \exp(i\theta)u$. Moreover, L is the vector bundle associated with P and the representation

$$\theta \in \mathbb{T} \mapsto (z \mapsto \exp(-i\theta)z) \in \mathrm{GL}(\mathbb{C})$$

of $\mathbb{T}$. There is a natural isomorphism of $\mathcal{C}^\infty(X)$-modules

$$\phi \colon \mathcal{C}^\infty(X, L) \to \{f \in \mathcal{C}^\infty(P) \mid R_\theta^* f = \exp(-i\theta)f\}, \quad s \mapsto f = \phi(s)$$

where, for $u \in P$, $f(u)$ is the unique complex number such that

$$s\big(\pi(u)\big) = f(u)u$$

where $\pi \colon P \to X$ is the canonical projection. Given any connection $\alpha \in \Omega^1(P)$ on P, we consider the connection ∇ on L such that the covariant derivative with respect to a vector field corresponds to the Lie derivative with respect to its horizontal lift:

$$\forall Y \in \mathcal{C}^\infty(X, TX), \forall s \in \mathcal{C}^\infty(X, L) \quad \phi(\nabla_Y s) = \mathcal{L}_{Y^{\mathrm{hor}}}\big(\phi(s)\big).$$

This map ∇ is well-defined because ϕ is an isomorphism, and it satisfies the Leibniz rule because the Lie derivative does and ϕ^{-1} is $\mathcal{C}^\infty(X)$-linear.

Exercise A.1.4. Carefully check all the above statements.

Lemma A.1.5. *The map sending α to ∇ is a bijection from the set of connections on P to the set of connections on L.*

Proof. Let us work with local trivialisations. Let $U \subset X$ be an open subset endowed with a unitary frame $s \in \mathcal{C}^\infty(U, L)$. We get a local trivialisation of P over U,

$$\varphi \colon P_{|U} \to U \times \mathbb{T}, \quad u \mapsto (\pi(u), \theta)$$

where θ is the unique element of $\mathbb{T}$ such that $s\big(\pi(u)\big) = \exp(i\theta)u$. Now, let us identify $\mathcal{C}^\infty(U, L)$ with $\mathcal{C}^\infty(U)$ by sending the section fs to f, and $\mathcal{C}^\infty(P_{|U})$ with $\mathcal{C}^\infty(U \times \mathbb{T})$ via φ. Then $\phi(f) = g$ with

$$g(x, \theta) = f(x)\exp(-i\theta).$$

Using these identifications, $\alpha = \beta + d\theta$ for some $\beta \in \Omega^1(U)$. Therefore, given some vector field Y on U, its horizontal lift is given by $Y^{\mathrm{hor}} = Y - \beta(Y)\partial_\theta$, hence

$$(\mathcal{L}_{Y^{\mathrm{hor}}} g)(x, \theta) = \left(d_x g(Y) - \beta(Y)\frac{\partial g}{\partial\theta}\right)(x, \theta) = (\mathcal{L}_Y f + i\beta(Y)f)(x)\exp(i\theta)$$

Consequently,

$$\nabla(fs) = (df + i\beta) \otimes s$$

so ∇ is uniquely determined by α. $\qquad\square$

A.2 The Szegő Projector of a Strictly Pseudoconvex Domain

Let Y be a complex manifold of complex dimension $n+1$. Let $D \subset Y$ be a domain (connected open subset) of Y with smooth compact boundary, defined as

$$D = \{y \in Y \mid \eta(y) < 0\}$$

with $\eta \colon Y \to \mathbb{R}$ smooth and such that $\mathrm{d}\eta(y) \neq 0$ whenever y belongs to ∂D. Let H be the complex subbundle of $T(\partial D) \otimes \mathbb{C}$ consisting of the holomorphic tangent vectors of Y which are tangent to the boundary of D; it has complex dimension n. The *Levi form* of D is the restriction to H of the quadratic form $\partial \bar{\partial} \eta$.

Definition A.2.1. We say that D is *strictly pseudoconvex* if its Levi form is positive definite at every point of ∂D.

Note that this implies that the restriction α of $-\mathrm{i}\partial\eta$ to ∂D is a contact form on ∂D. Thus we get a volume form $\mu = \alpha \wedge (\mathrm{d}\alpha)^n$ on ∂D, and we can consider the Hilbert space $L^2(\partial D)$ with respect to μ. The subspace

$$\mathcal{H}(D) = \{f \in L^2(\partial D) \mid \forall Z \in \mathcal{C}^\infty(\partial D, H)\, \mathcal{L}_{\bar{Z}} f = 0\}$$

is called the *Hardy space* of D. The *Szegő projector* of D is the orthogonal projector $\Pi \colon L^2(\partial D) \to \mathcal{H}(D)$.

A.3 Application to Geometric Quantisation

Coming back to our problem, where M is a compact Kähler manifold and $L \to M$ is a prequantum line bundle, let us introduce the $\mathbb{T}$-principal bundle $P \to M$ which consists of unit norm elements (with respect to the norm induced by h) of the line bundle L. It is such that for every integer k, we have the line bundle isomorphism $L^k \simeq P \times_{s_k} \mathbb{C}$ where $s_k \colon \mathbb{T} \to \mathrm{GL}(\mathbb{C})$ is the representation given by

$$s_k(\theta) \cdot v = \exp(-\mathrm{i}k\theta)v$$

We can embed P into $L^{-1} \simeq P \times_{s_{-1}} \mathbb{C}$ via

$$\iota \colon P \to P \times_{s_{-1}} \mathbb{C}, \quad \iota(p) = [p, 1]$$

where the square brackets stand for equivalence class. The connection on L^{-1}, that we still denote by ∇, induces a connection one-form $\alpha \in \Omega^1(P)$. Let $\mathrm{Hor}^{1,0}$ be the subbundle of $TP \otimes \mathbb{C}$ consisting of the horizontal lifts of the holomorphic vectors of $TM \otimes \mathbb{C}$. Let

$$\rho \colon L^{-1} \to \mathbb{R}, \quad u \mapsto \|u\|^2$$

and let $D = \{u \in L^{-1} \mid \rho(u) < 1\}$.

Proposition A.3.1. *D is a strictly pseudoconvex domain of L^{-1} and $\partial D = \iota(P)$. The bundle H of holomorphic vectors of L^{-1} that are tangent to $\iota(P)$ is $\iota_* \mathrm{Hor}^{1,0}$. Moreover, $\iota^* \partial \log \rho = i\alpha$.*

Proof. We begin by proving the second assertion. Let us use some local coordinates. Let $U \subset M$ be an open subset such that $P_{|U} \simeq U \times \mathbb{T}$, and let us use coordinates (x, θ) on $U \times \mathbb{T}$. Then $\alpha = \beta + d\theta$ for some $\beta \in \Omega^1(U)$. Let s^{-1} be the local section of $L^{-1} \to U$ defined by

$$s^{-1}(x) = [(x, 0), 1] \in (U \times \mathbb{T}) \times_{s^{-1}} \mathbb{C} \simeq L_{|U}^{-1}.$$

Then $\nabla s^{-1} = i\beta \otimes s^{-1}$. We pick a function $\phi \in \mathcal{C}^\infty(U)$ such that

$$\bar\partial \phi + i\beta^{(0,1)} = 0; \tag{A.1}$$

we know that such a function exists (taking a smaller U if necessary) thanks to the Dolbeault–Grothendieck lemma, since $d\beta$ is a $(1,1)$-form. Then

$$\nabla(\exp(\phi)s^{-1}) = \exp(\phi)(\partial\phi + \bar\partial\phi + i\beta) \otimes s^{-1} = \exp(\phi)(\partial\phi + i\beta^{(1,0)}) \otimes s^{-1}$$

hence $\exp(\phi)s^{-1}$ is a holomorphic section. Let w be the complex linear coordinate of L^{-1} such that $w(\exp(\phi)s^{-1}) = 1$, and let $(z_j)_{1 \leq j \leq n}$ be a system of complex coordinates on U. In these coordinates, the maps ι and ρ read

$$\iota: U \times \mathbb{T} \to U \times \mathbb{C}, \quad (z_1, \ldots, z_n, \theta) \mapsto \left(z_1, \ldots, z_n, w = \exp\big(i\theta - \phi(z)\big)\right)$$

and

$$\rho: U \times \mathbb{C} \to \mathbb{R}, \quad (z_1, \ldots, z_n, w) \mapsto |w|^2 \exp\big(\phi(z) + \bar\phi(z)\big).$$

Let $j \in [\![1, n]\!]$; the horizontal lift of ∂_{z_j} is

$$\partial_{z_j}^{\mathrm{hor}} = \partial_{z_j} - \beta(\partial_{z_j})\partial_\theta$$

We compute

$$\beta(\partial_{z_j}) = \beta^{(1,0)}(\partial_{z_j}) = -i\frac{\partial\bar\phi}{\partial z_j},$$

the last equality coming from the fact that $\partial\bar\phi - i\beta^{(1,0)} = 0$ because β is real-valued and satisfies (A.1). Hence

$$\partial_{z_j}^{\mathrm{hor}} = \partial_{z_j} + i\frac{\partial\bar\phi}{\partial z_j}\partial_\theta.$$

Therefore, its pushforward by ι satisfies

$$\iota_*\big(\partial_{z_j}^{\mathrm{hor}}\big) = dz_j\left(\partial_{z_j} + i\frac{\partial\bar\phi}{\partial z_j}\partial_\theta\right)\partial_{z_j} + dw\left(\partial_{z_j} + i\frac{\partial\bar\phi}{\partial z_j}\partial_\theta\right)\partial_w,$$

which yields

$$\iota_*\left(\partial_{z_j}^{\mathrm{hor}}\right) = \partial_{z_j} + \mathrm{d}w\left(\partial_{z_j} + \mathrm{i}\frac{\partial\bar\phi}{\partial z_j}\partial_\theta\right)\partial_w.$$

Since $\mathrm{d}w = w(\mathrm{i}\mathrm{d}\theta - \mathrm{d}\phi)$, we finally obtain that

$$\iota_*\left(\partial_{z_j}^{\mathrm{hor}}\right) = \partial_{z_j} - \frac{\partial(\phi + \bar\phi)}{\partial z_j}\partial_w.$$

This implies that $\iota_*\mathrm{Hor}^{1,0}$ is a subbundle of the bundle H of holomorphic vectors of L^{-1} which are tangent to $\iota(P)$; since both bundles have complex dimension n, this means that they are equal.

Let us now prove the last claim of the proposition. We have that

$$\partial\rho = \exp(\phi + \bar\phi)\left(\overline{w}\,\mathrm{d}w + |w|^2\partial(\phi + \bar\phi)\right),$$

hence

$$\partial(\log\rho) = \frac{\mathrm{d}w}{w} + \partial(\phi + \bar\phi).$$

Consequently,

$$\iota^*\partial(\log\rho) = \mathrm{i}\mathrm{d}\theta - \mathrm{d}\phi + \partial(\phi + \bar\phi) = \mathrm{i}\mathrm{d}\theta - \bar\partial\phi + \partial\bar\phi.$$

Remembering (A.1) and the conjugate equality, we finally obtain that

$$\iota^*\partial(\log\rho) = \mathrm{i}(\mathrm{d}\theta + \beta) = \mathrm{i}\alpha.$$

It remains to show that D is strictly pseudoconvex. Its Levi form is equal to the restriction of $\iota^*(\partial\bar\partial\log\rho)$ to $H = \iota_*\mathrm{Hor}^{1,0}$. But

$$\iota^*(\partial\bar\partial\log\rho) = -\iota^*(\bar\partial\partial\log\rho) = -\iota^*(\mathrm{d}\bar\partial\log\rho) = -\,\mathrm{d}\iota^*(\partial\log\rho) = -\mathrm{i}\mathrm{d}\alpha.$$

Since $-\mathrm{i}\mathrm{d}\alpha$ corresponds to the curvature of the connection on L over U, we have that

$$-\mathrm{i}\mathrm{d}\alpha\left(\partial_{z_j}^{\mathrm{hor}}, \partial_{\bar z_\ell}^{\mathrm{hor}}\right) = -\mathrm{i}\omega(\partial_{z_j}, \partial_{\bar z_\ell}) > 0,$$

which concludes the proof. □

As a consequence of this result, we construct the Hilbert space $L^2(P)$ by using the volume form $\mu_P = (1/(2\pi n!))\alpha \wedge (\mathrm{d}\alpha)^n$, the Hardy space

$$\mathcal{H}(P) = \{f \in L^2(P) \,|\, \forall Z \in \mathcal{C}^\infty(P, H), \mathcal{L}_{\bar Z}f = 0\} \subset L^2(P)$$

as in the previous section and the Szegő projector $\Pi\colon L^2(P) \to \mathcal{H}(P)$.

Since $L^k \simeq P \times_{s_k} \mathbb{C}$, we have an identification

$$\mathcal{C}^\infty(M, L^k) \to \{f \in \mathcal{C}^\infty(P) \,|\, R_\theta^* f = \exp(\mathrm{i}k\theta)f\}$$

which sends $s \in \mathcal{C}^\infty(M, L^k)$ to $f \in \mathcal{C}^\infty(P)$, where, for $p \in P$, $f(p)$ is the unique complex number such that $s(\pi(p)) = f(p)p$.

Lemma A.3.2. *This identification is compatible with the scalar products on $\mathcal{C}^\infty(P)$ and $\mathcal{C}^\infty(M, L^k)$ (i.e., it defines an isometry).*

Proof. Let $s, t \in \mathcal{C}^\infty(M, L^k)$ and let $f, g \in \mathcal{C}^\infty(P)$ be the corresponding functions. Observe that for $p \in P$,

$$h_k\Big(s(\pi(p)), t(\pi(p))\Big) = f(p)\bar{g}(p)$$

since $h(p, p) = 1$. Therefore, we have that

$$\langle f, g \rangle_P = \int_P f\bar{g}\, \mu_P = \int_P \pi^*\big(h_k(s, t)\big)\, \mu_P.$$

Since $\alpha \wedge (\mathrm{d}\alpha)^n = \mathrm{d}\theta \wedge \pi^*\omega^n$, we deduce from this equality that

$$\langle f, g \rangle_P = \int_M h_k(s, t)\, \mu = \langle s, t \rangle_k,$$

which was to be proved. □

Under this identification, the covariant derivative $\nabla_X s$ corresponds to the Lie derivative $\mathcal{L}_{X^{\mathrm{hor}}} f$; hence, s is holomorphic if and only if f belongs to $\mathcal{H}(P)$, since, as we saw earlier, $H = \iota_* \mathrm{Hor}^{1,0}$. By Fourier decomposition, we have the splitting

$$L^2(P) = \bigoplus_{k \in \mathbb{Z}} \{f \in L^2(P) \mid \forall \theta \in \mathbb{T}, R_\theta^* f = \exp(\mathrm{i}k\theta)f\}.$$

To be more precise, $(R_\theta^*)_{\theta \in \mathbb{T}}$ is a family of commuting self-adjoint operators acting on $L^2(P)$, each R_θ^* has discrete spectrum $\big(\exp(\mathrm{i}k\theta)\big)_{k \in \mathbb{Z}}$, therefore they all have the same eigenspaces, and $L^2(P)$ splits into the direct sum of these eigenspaces. Now, using the above lemma, this yields a unitary isomorphism

$$L^2(P) \simeq \bigoplus_{k \in \mathbb{Z}} L^2(M, L^k).$$

Since Π commutes with every R_θ^*, $\theta \in \mathbb{T}$, we also obtain the unitary equivalence

$$\mathcal{H}(P) \simeq \bigoplus_{k \in \mathbb{Z}} H^0(M, L^k) = \bigoplus_{k \in \mathbb{Z}} \mathcal{H}_k = \bigoplus_{k \geq 0} \mathcal{H}_k,$$

where the last equality comes from Proposition 4.2.1, and Π_k corresponds to the Fourier coefficient at order k of Π, that is its restriction to the space $L^2(M, L^k)$.

One can use this approach to derive another proof of Theorem 7.2.1, in a way that we quickly describe now. In their seminal article [34], Boutet de Monvel and Sjöstrand obtained a precise description of the Schwartz kernel of this projector, that we describe now. Let $\phi \in \mathcal{C}^\infty(Y \times Y)$ be such that

$$\phi(y,y) = -i\eta, \quad \phi(x,y) = -\overline{\phi(y,x)}, \quad \mathcal{L}_{\overline{Z}_\ell}\phi \equiv \mathcal{L}_{Z_r}\phi \equiv 0 \bmod \mathcal{I}^\infty(\mathrm{diag}(Y^2))$$

for every holomorphic vector field Z, where Z_ℓ (respectively Z_r) means acting on the left (respectively right) variable, and $\mathcal{I}^\infty(\mathrm{diag}(Y^2))$ is the set of functions vanishing to infinite order along the diagonal of Y^2. It is known that such a function ϕ exists and is unique up to a function vanishing to infinite order along the diagonal of Y^2.

Define $\varphi \in \mathcal{C}^\infty(\partial D \times \partial D)$ as the restriction of ϕ to $\partial D \times \partial D$. Then $\mathrm{d}\varphi$ does not vanish on $\mathrm{diag}(\partial D \times \partial D)$, whereas $\mathrm{d}(\Im\,\varphi)$ vanishes on $\mathrm{diag}(\partial D \times \partial D)$ and has negative Hessian with kernel $\mathrm{diag}(T\partial D \times T\partial D)$. Thus we may assume, by modifying φ outside a neighbourhood of $\mathrm{diag}(\partial D \times \partial D)$ if necessary, that $\Im\,\varphi(u_\ell, u_r) < 0$ if $u_\ell \neq u_r$.

Theorem A.3.3. *([34, Theorem 1.5]) The Schwartz kernel of the Szegő projector Π satisfies*

$$\Pi(u_\ell, u_r) = \int_{\mathbb{R}^+} \exp(i\tau\varphi(u_\ell, u_r))\, s(u_\ell, u_r, \tau)\, \mathrm{d}\tau + f(u_\ell, u_r)$$

where $f \in \mathcal{C}^\infty(\partial D \times \partial D)$ and $s \in S^n(\partial D \times \partial D \times \mathbb{R}^+)$ is a classical symbol having the asymptotic expansion

$$s(u_\ell, u_r, \tau) \sim \sum_{j \geq 0} \tau^{n-j} s_j(u_\ell, u_r).$$

Theorem 7.2.1 can be inferred from this result, the idea being that one can deduce the asymptotics of Π_k when k goes to infinity from the description of the Schwartz kernel of Π, in a way which is similar to the deduction of the behaviour of the Fourier coefficients of a function at $\pm\infty$ from the regularity of this function. For a detailed proof using this approach, one can, for example, look at Section 3.3 in [14].

References

1. Andersen, J.E., Blaavand, J.L.: Asymptotics of Toeplitz operators and applications in TQFT. In: M. Schlichenmaier, A. Sergeev, O. Sheinman (eds.) Geometry and Quantization, *Trav. Math.*, vol. 19, pp. 167–201. Univ. of Luxembourg, Luxembourg (2011)
2. Bargmann, V.: On a Hilbert space of analytic functions and an associated integral transform. Comm. Pure Appl. Math. **14**, 187–214 (1961)
3. Bargmann, V.: On a Hilbert space of analytic functions and an associated integral transform. Part II. A family of related function spaces. Application to distribution theory. Comm. Pure Appl. Math. **20**, 1–101 (1967)
4. Barron, T., Ma, X., Marinescu, G., Pinsonnault, M.: Semi-classical properties of Berezin–Toeplitz operators with $\mathcal{C}^k$-symbol. J. Math. Phys. **55**(4), 042108, 25 pp. (2014). DOI https://doi.org/10.1063/1.4870869
5. Berezin, F.A.: General concept of quantization. Comm. Math. Phys. **40**, 153–174 (1975)
6. Berman, R., Berndtsson, B., Sjöstrand, J.: A direct approach to Bergman kernel asymptotics for positive line bundles. Ark. Mat. **46**(2), 197–217 (2008). https://doi.org/10.1007/s11512-008-0077-x
7. Bloch, A., Golse, F., Paul, T., Uribe, A.: Dispersionless Toda and Toeplitz operators. Duke Math. J. **117**(1), 157–196 (2003). https://doi.org/10.1215/S0012-7094-03-11713-5
8. Bordemann, M., Meinrenken, E., Schlichenmaier, M.: Toeplitz quantization of Kähler manifolds and gl(N), $N \to \infty$ limits. Comm. Math. Phys. **165**(2), 281–296 (1994)
9. Borthwick, D., Paul, T., Uribe, A.: Semiclassical spectral estimates for Toeplitz operators. Ann. Inst. Fourier (Grenoble) **48**(4), 1189–1229 (1998)
10. Borthwick, D., Uribe, A.: Almost complex structures and geometric quantization. Math. Res. Lett. **3**(6), 845–861 (1996). https://doi.org/10.4310/MRL.1996.v3.n6.a12
11. Bott, R., Tu, L.W.: Differential Forms in Algebraic Topology, *Grad. Texts in Math.*, vol. 82. Springer, New York (1982)
12. Brylinski, J.-L.: Loop Spaces, Characteristic Classes and Geometric Quantization. Modern Birkhäuser Classics. Birkhäuser, Boston, MA (2008). DOI https://doi.org/10.1007/978-0-8176-4731-5
13. Cahen, M., Gutt, S., Rawnsley, J.: Quantization of Kähler manifolds. II. Trans. Amer. Math. Soc. **337**(1), 73–98 (1993). https://doi.org/10.2307/2154310
14. Charles, L.: Berezin-Toeplitz operators, a semi-classical approach. Comm. Math. Phys. **239**(1–2), 1–28 (2003). https://doi.org/10.1007/s00220-003-0882-9
15. Charles, L.: Quasimodes and Bohr-Sommerfeld conditions for the Toeplitz operators. Comm. Partial Differential Equations **28**(9–10), 1527–1566 (2003). https://doi.org/10.1081/PDE-120024521
16. Charles, L.: Symbolic calculus for Toeplitz operators with half-form. J. Symplectic Geom. **4**(2), 171–198 (2006)

© Springer International Publishing AG, part of Springer Nature 2018
Y. Le Floch, *A Brief Introduction to Berezin–Toeplitz Operators on Compact Kähler Manifolds*, CRM Short Courses,
https://doi.org/10.1007/978-3-319-94682-5

17. Charles, L.: Quantization of compact symplectic manifolds. J. Geom. Anal. **26**(4), 2664–2710 (2016). https://doi.org/10.1007/s12220-015-9644-0

18. Charles, L., Marché, J.: Knot state asymptotics. I. AJ conjecture and Abelian representations. Publ. Math., Inst. Hautes Étud. Sci. **121**, 279–322 (2015). DOI https://doi.org/10.1007/s10240-015-0068-y

19. Charles, L., Marché, J.: Knot state asymptotics. II. Witten conjecture and irreducible representations. Publ. Math., Inst. Hautes Étud. Sci. **121**, 323–361 (2015). DOI https://doi.org/10.1007/s10240-015-0069-x

20. Charles, L., Polterovich, L.: Sharp correspondence principle and quantum measurements. Algebra i Analiz 29(1), 237–278 (2017). St. Petersburg Math. J. **29**(1), 177–207 (2018). https://doi.org/10.1090/spmj/1488

21. Charles, L., Polterovich, L.: Quantum speed limit versus classical displacement energy. Ann. Henri Poincaré **19**(4), 1215–1257 (2018). https://doi.org/10.1007/s00023-018-0649-7

22. Donaldson, S.K.: Scalar curvature and projective embeddings. I. J. Differential Geom. **59**(3), 479–522 (2001)

23. Guillemin, V.: Star products on compact pre-quantizable symplectic manifolds. Lett. Math. Phys. **35**(1), 85–89 (1995). https://doi.org/10.1007/BF00739157

24. Huybrechts, D.: Complex Geometry. An Introduction. Universitext. Springer, Berlin (2005). DOI https://doi.org/10.1007/b137952

25. Hörmander, L.: An Introduction to Complex Analysis in Several Variables, *North-Holland Math. Library*, vol. 7. Elsevier, Amsterdam (1973)

26. Hörmander, L.: The Analysis of Linear Partial Differential Operators. I. Distribution Theory and Fourier Analysis, *Grundlehren Math. Wiss.*, vol. 256. Springer, Berlin (1983). DOI https://doi.org/10.1007/978-3-642-96750-4

27. Karabegov, A.V., Schlichenmaier, M.: Identification of Berezin-Toeplitz deformation quantization. J. Reine Angew. Math. **540**, 49–76 (2001). https://doi.org/10.1515/crll.2001.086

28. Kostant, B.: Quantization and unitary representations. I. Prequantization. In: Lectures in Modern Analysis and Applications. III, *Lecture Notes in Math.*, vol. 170, pp. 87–208. Springer, Berlin (1970). Russian transl., Uspehi Mat. Nauk **28**(1), 163–225 (1973)

29. Ma, X., Marinescu, G.: Holomorphic Morse Inequalities and Bergman Kernels, *Progr. Math.*, vol. 254. Birkhäuser, Basel (2007)

30. Ma, X., Marinescu, G.: Toeplitz operators on symplectic manifolds. J. Geom. Anal. **18**(2), 565–611 (2008). https://doi.org/10.1007/s12220-008-9022-2

31. Ma, X., Marinescu, G.: Berezin-Toeplitz quantization on Kähler manifolds. J. Reine Angew. Math. **662**, 1–56 (2012). https://doi.org/10.1515/CRELLE.2011.133

32. Marché, J., Paul, T.: Toeplitz operators in TQFT via skein theory. Trans. Am. Math. Soc. **367**(5), 3669–3704 (2015)

33. Boutet de Monvel, L., Guillemin, V.: The Spectral Theory of Toeplitz Operators, *Ann. of Math. Stud.*, vol. 99. Princeton Univ. Press, Princeton, NJ (1981)

34. Boutet de Monvel, L., Sjöstrand, J.: Sur la singularité des noyaux de Bergman et de Szegö. In: J. Camus (ed.) Journées: Équations aux Dérivées Partielles de Rennes (1975), no. 34-35 in Astérisque, pp. 123–164. Soc. Math. France, Paris (1976)

35. Moroianu, A.: Lectures on Kähler Geometry, *London Math. Soc. Stud. Texts*, vol. 69. Cambridge Univ. Press, Cambridge (2007)

36. Mumford, D.: Tata Lectures on Theta. I. Modern Birkhäuser Classics. Birkhäuser, Boston, MA (2007). DOI https://doi.org/10.1007/978-0-8176-4578-6. With the collaboration of C. Musili, M. Nori, E. Previato and M. Stillman

37. Polterovich, L.: Quantum unsharpness and symplectic rigidity. Lett. Math. Phys. **102**(3), 245–264 (2012). https://doi.org/10.1007/s11005-012-0564-7

38. Polterovich, L.: Symplectic geometry of quantum noise. Comm. Math. Phys. **327**(2), 481–519 (2014). https://doi.org/10.1007/s00220-014-1937-9

39. Rawnsley, J., Cahen, M., Gutt, S.: Quantization of Kähler manifolds. I. Geometric interpretation of Berezin's quantization. J. Geom. Phys. **7**(1), 45–62 (1990). https://doi.org/10.1016/0393-0440(90)90019-Y

40. Rawnsley, J.H.: Coherent states and Kähler manifolds. Quart. J. Math. Oxford Ser. (2) **28**(112), 403–415 (1977). DOI https://doi.org/10.1093/qmath/28.4.403
41. Rubinstein, Y.A., Zelditch, S.: The Cauchy problem for the homogeneous Monge-Ampère equation. I. Toeplitz quantization. J. Differential Geom. **90**(2), 303–327 (2012)
42. Schlichenmaier, M.: Deformation quantization of compact Kähler manifolds by Berezin–Toeplitz quantization. In: G. Dito, D. Sternheimer (eds.) Conférence Moshé Flato 1999, Vol. II (Dijon, 1999), *Math. Phys. Stud.*, vol. 22, pp. 289–306. Kluwer, Dordrecht (2000)
43. Schlichenmaier, M.: Berezin–Toeplitz quantization for compact Kähler manifolds. A review of results. Adv. Math. Phys. **2010**, 927280, 38 pp. (2010)
44. Shiffman, B., Zelditch, S.: Asymptotics of almost holomorphic sections of ample line bundles on symplectic manifolds. J. Reine Angew. Math. **544**, 181–222 (2002). https://doi.org/10. 1515/crll.2002.023
45. Souriau, J.-M.: Quantification géométrique. Comm. Math. Phys. **1**, 374–398 (1966)
46. Tuynman, G.M.: Quantization: Towards a comparison between methods. J. Math. Phys. **28**(12), 2829–2840 (1987). https://doi.org/10.1063/1.527681
47. Woodhouse, N.M.J.: Geometric Quantization, 2nd edn. Oxford Math. Monogr. Oxford Univ. Press, New York (1992)
48. Zelditch, S.: Index and dynamics of quantized contact transformations. Ann. Inst. Fourier (Grenoble) **47**(1), 305–363 (1997)
49. Zelditch, S.: Szegö kernels and a theorem of Tian. Internat. Math. Res. Notices **1998**(6), 317–331 (1998). https://doi.org/10.1155/S107379289800021X

Index of Notations

© Springer International Publishing AG, part of Springer Nature 2018
Y. Le Floch, *A Brief Introduction to Berezin–Toeplitz Operators on Compact Kähler Manifolds*, CRM Short Courses,
https://doi.org/10.1007/978-3-319-94682-5

Index

© Springer International Publishing AG, part of Springer Nature 2018
Y. Le Floch, *A Brief Introduction to Berezin–Toeplitz Operators on Compact Kähler Manifolds*, CRM Short Courses,
https://doi.org/10.1007/978-3-319-94682-5

Printed in the United States
By Bookmasters